The Search for Life on Mars

The Search for Life on Mars

Malcolm Walter

Foreword by Paul Davies

PERSEUS BOOKS
Cambridge, Massachusetts

Copyright © 1999 by Malcolm Walter

Published in Australia by Allen & Unwin Pty, Ltd.

Library of Congress Catalog Card Number: 99-65855
ISBN: 0-7382-0124-3

Perseus Books is a member of the Perseus Books Group

Find us on the World Wide Web at http://www.perseusbooks.com

Perseus Books are available at special discounts for bulk purchases in the U.S. by corporations, institutions, and other organizations. For more information, please contact the Special Markets Department at HarperCollins Publishers, 10 East 53rd Street, New York, NY 10022, or call 1–212–207–7528.

1 2 3 4 5 6 7 8 9 10—03 02 01 00 99
First printing, August 1999

To
my children
Amanda, Julian
and Serena

May
they live to learn
whether life is a
Cosmic imperative

This is in some measure an act of faith and vision, for we do not know what benefits await us . . .

John F. Kennedy, of going to the Moon

CONTENTS

Acknowledgements

Paul Davies suggested that I write this book. I thank him for that, and for what I have learned. Ian Bowring is my publisher; his enthusiasm, tolerance and finally his admonitions are now manifest.

Susie Rix is my wife, my most trenchant critic, my greatest supporter, and a bright window on a life not driven by science; no thanks can be sufficient.

David DesMarais first invited me to become involved with NASA's Mars program, contributing to a planning meeting in California in 1987; he probably does not realise how much I owe him. That invitation has proven to be an epiphany. With the openness characteristic of American science, he and my other NASA colleagues have given me the greatest opportunity of my scientific career. Furthermore, over the last decade, a substantial amount of my research has been funded by NASA. The Australian Research Council and Macquarie University in Sydney have also supported me. I am deeply grateful.

Many of the things I have written about lie outside my own area of direct knowledge. My colleagues have helped me with advice and information, and by critically reading part or all of the book: Paul Davies, David DesMarais, Kath Grey, Andrew Knoll, John Lindsay, Reg Morrison, Michael Russell, Duncan Steel, Roger Summons, Ross Taylor.

My friends, colleagues and students at Macquarie University have given me an intellectual context, and their support, without which my obsession with Mars would have withered. Foremost among them is John Veevers.

This book is a product of the more than 35 years' experience of science that now lies behind me. I have learned something valuable from everyone I have known along the way. My sixteen years with the Australian Bureau of Mineral Resources taught me great respect for their work. The greatest experience and privilege of all during that time was to lead the Baas Becking Geobiological Laboratory, an organisation before its time and thus, sadly, no longer existing. During those sixteen years I was allowed time out to work in the Precambrian Paleobiology Research Group at the University of California in Los Angeles, under the exacting leadership of Bill Schopf, another defining event in my career. These two experiences taught me the skill of interdisciplinary research, on which any search for life elsewhere must be based.

Many people read parts or all of the book at some stage, and tried to help me. They will find themselves acknowledged in one way or another. Particular thanks go to my editors Emma Cotter and Karen Ward.

Perhaps it is vain to think that someone will read this book and care. But maybe they will. I have bared a little of my soul.

FOREWORD

Mars has always held a particular fascination for people, partly because of the rich red colour that gives it such prominence in the night sky, but also because of the possibility that it harbours life. The ancients were convinced that the Red Planet was inhabited, a view that prevailed through the Middle Ages right up to the modern scientific era. In the 1890s the American astronomer Percival Lowell suggested that intelligent beings had built a network of canals there, and the science fiction writer H.G. Wells brilliantly exploited popular fears by choosing Martians as the invading force in his novel *War of the Worlds*. Although with improved scientific understanding Lowell's claim came to seem incredible, even as late as the 1960s it was widely supposed that at least some sort of vegetation grew on Mars.

When the first space probes went to the planet in the 1960s, they did not lend much support to the possibility of life. Mars was revealed as a freeze-dried desert with a pitifully thin atmosphere, bathed in deadly ultra-violet rays and exposed to cosmic radiation. Hopes that primitive microbes might yet flourish on the Martian surface were dashed when, in 1976, two *Viking Lander* space probes tested the topsoil for organisms—without success.

For twenty years most scientists wrote off the Red Planet as an abode for life. But then the mood began to change. Surveys showed that although the surface of Mars is today highly inhospitable, in the remote past conditions there were warm and wet, not unlike those on Earth. Certainly the atmosphere was thicker then, and liquid water sometimes ran freely, producing river valleys, flood plains and possibly even an ocean. Second, it was discovered that terrestrial life is able to thrive in the most extreme conditions. Microbes have been found both in the searing water that emerges from deep ocean springs near volcanic vents and kilometres underground, at temperatures that exceed 100°C. Significantly, Mars also has volcanoes and the combination of water and volcanic activity must have created hydrothermal systems on Mars very similar to those on Earth—where life abounds. It no longer seems unreasonable to believe that there was microbial life on Mars some billions of years ago. It may even be that colonies of microorganisms still cling on deep below the surface, where geothermal heat has melted the permafrost.

The subject of life on Mars was propelled into prominence three years ago when the US space agency NASA announced it had evidence for Martian microbes in the form of tiny features in a meteorite found in Antarctica, known to have come from Mars. Although subsequent tests have not confirmed the claim, the huge attention that the meteorite received provided a welcome fillip for research into extraterrestrial life, and led to an improvement in laboratory analyses that should greatly benefit future Mars exploration.

Recognising the interest among the public and scientific community alike in the search for life beyond Earth, NASA recently launched its astrobiology program, with Mars being a prime objective. Over the coming years, the Red Planet will be explored by a veritable armada of space probes, many of which could cast light on the possibility that life once existed on the surface.

Unfortunately, any attempt to answer the question, 'Was there life on Mars?' is beset with difficulties. Since it is extremely unlikely that any Martian organisms ever progressed beyond the level of microbes, the problem is whether any trace of past life remains on the surface of the planet. Most experts agree that the warm and wet period on Mars probably ended about 3.5 billion years ago (with possible short episodes since), so any relic of past life would have to be extraordinarily well preserved. Would Martian microbes have left fossils in rocks, or other biomarkers that would reveal their existence such a long time after? If so, where exactly should scientists look to find these traces?

To try and answer these questions, it makes sense to start with the record of ancient life on Earth. Are there remnants of terrestrial microorganisms that lived 3.5 billion years ago? Malcolm Walter is a world expert on palaeobiology, in particular the study of fossilised microbes from the earliest geological epochs. He was closely involved in the discovery of the world's oldest fossils, found in the Pilbara region of Western Australia. He is a NASA consultant on astrobiology and a leading proponent of the theory that Earth's earliest organisms dwelt around hydrothermal systems; indeed, it is looking increasingly likely that life was originally incubated in such a setting. Since Mars almost certainly had similar hydrothermal systems to Earth 3.5 billion years ago, Walter argues that their remains would offer the best chance for finding fossilised Martians.

This fascinating book provides a valuable window on the exhilarating, frustrating and occasionally acrimonious world of palaeobiology and astrobiology. As such, it is representative of all science. Many lay people fondly imagine that scientists go about their work in a cold, calculating and objective manner, sure of their facts and meticulous in their theories. The reality is very different.

Science is a passionate, sometimes intuitive enterprise, riven by competing theories, strong egos and warring factions. Evidence is often ambiguous or incomplete, great ideas can be framed on mere hunches and fashion may dictate what gets investigated and what is ignored. Funding, or rather the lack of it, can dominate research trends. Instead of presenting a cosmeticised view, Walter takes us on a blow-by-blow account of the convoluted and at times grubby history of his subject, portraying the key investigators as real people, with real ambitions and rivalries. He skilfully conveys how hard it can be to interpret correctly the record of the rocks, and just what a huge challenge scientists face in their attempt to identify life on another world. Anyone who thinks that it is a simple matter to send a probe to Mars, grab a few rocks, bring them back to Earth and look for fossils is being very optimistic.

In spite of the problem that astrobiology faces, Walter never loses sight of the big picture—the magnificence of mankind's exploration of the cosmos. He regards the project to seek out life on Mars as the first step on the road to humanity's cosmic destiny, leading to colonisation of the Red Planet and the search for life on more distant moons and planets. This is a truly inspirational account of the greatest adventure in history, told by one of its most influential players.

Paul Davies
Adelaide, 1999

CHAPTER 1

EARTH, THE MOON, PEOPLE, MARS, THE UNIVERSE

There will be people on Mars long before the end of the twenty-first century. It's inevitable, and irresistible. It might happen before 2020. It could happen by 2011. Mars is our next frontier. The plans are being laid now, the missions designed. The technology exists. The latter-day equivalents of Magellan, Columbus and Cook, and all the other explorers of the age of European expansion, are preparing themselves. The motives are many, as always, but central for scientists is the search for life, or former life. At stake is the issue, are we alone in the Universe?

In October 1957, at the age of thirteen and like everyone else in my street, I went out in the dark, with my parents, to watch the first satellite, *Sputnik*, orbiting overhead. In September 1959, the Soviets achieved the first unmanned landing on the Moon. In April 1961, Yuri Gagarin became the first human to orbit the Earth. American astronaut John Glenn followed in 1962, and by then the space race was in full cry. The United States regained the initiative when on 20 July 1969 Neil Armstrong stepped from the *Apollo 11* lander *Eagle* at Tranquillity Base on the Moon, 'one small step for man, one giant leap for Mankind'. By then I was more than old enough to sit up all night drinking beer and playing cards with friends while we watched the landing on television and waited for Armstrong and Edwin ('Buzz') Aldrin to step out.

The Moon was not the only prize in the race. The exploration of Mars had also been a goal of the Russian and American space programs from the beginning. The Russians achieved the first flyby with *Mars 1* which was launched in November 1962. *Mariner 4* from the United States took the first pictures of Mars in 1965. The search for life on Mars was the primary purpose of *Viking*s *1* and *2* from the United States, which landed in 1976. In that year I was well into my career as a palaeontologist, but I certainly never even imagined that, starting in 1987, I would be involved in the exploration of Mars.

This book is my personal view of the search for the origin of life. Much of the appeal of science for me, and for most other scientists, is the excitement of discovery and understanding. We strive to be objective, but objectivity is close to impossible in the field of science I am writing about. There is a potent mix of sketchy knowledge, huge media and public interest, and a small number of practitioners, some with large and fragile egos (mine is one of them). To make matters worse, we are working in a social context that makes achieving objectivity even harder than usual: just about everyone wants to believe that there is life elsewhere, and I expect all involved scientists sense this collective wish. As American evolutionary biologist and author Stephen J. Gould has written, 'The more important the subject and the closer it cuts to the bone of our hopes and needs, the more we are likely to err in establishing a framework for analysis'. It is impossible to understand the search for earliest life on Earth and Mars without taking account of public perceptions and the personalities of the participating scientists. Science itself makes no sense except as part of culture, which, all in all, is a messy business. The striving for objectivity is real, but the reality is often something different. The scientific method is very effective in revealing and countering subjectivity or bias, but it is a bit like going to a wonderful restaurant: perhaps it is better to not know what is happening in the kitchen. But I can't resist giving you a peek. The result is a book that is unashamedly and explicitly my personal view of reality, and idiosyncratic.

I have set out here to communicate the insights and discoveries of scientists in this field, our excitement and our arguments. Your understanding should be guarded by an attitude of healthy scepticism. As stated by biologist Jared Diamond, among others, the field of science dealing with life elsewhere in the Universe, a field termed 'exobiology', or more recently, 'astrobiology', 'is the sole scientific field whose subject matter has not yet been shown to exist'. I could quibble with this and suggest that theoretical scientists often work with objects or processes that are inferred but not observed. Anyway, I write as an optimist, and expect us to learn a great deal from our search. This is also a cautionary tale, because many of the mistakes we have made in revealing the early history of life on Earth seem set to be repeated for Mars.

In searching for evidence of life on Mars we have only one example to work with: Earth. We can speculate about forms of life completely unfamiliar to us, maybe based on elements other than carbon, and we should stay alert for all the possibilities we can imagine. But all current searches of which I am aware take the more conservative approach of prediction based on what we know about Earth. The property of carbon to be able to form an enormous diversity of compounds is central to its role in living organisms. The search for life on Mars is based on the assumptions that if there is or was ever life there it would be carbon-based and require water. Not only that, in its simplest single-celled form it may well resemble similar life on Earth.

The SETI program, that is, the Search for Extra-Terrestrial Intelligence, which searches for radio signals from other civilisations, is another way to look for life elsewhere. Radio telescopes in several countries are being used in this search. SETI has to contend with enormous odds, given the vastness of the Universe and the uncertainties of recognising biogenic signals (those from living organisms). The odds may well be more in favour of discovering whether there once was, and maybe still is, life on Mars. SETI is like a fishing expedition where the quarry is unknown and the appropriate bait can only be guessed at. In contrast, early in their history, Earth and Mars were similar, allowing us to make the simple assumption that we are looking for something familiar: we will recognise the quarry and know how to bait the line. What is the chance of success? Well, if there was life, and it had a familiar form, eventually we will find the evidence. As I discuss in this book, there are good reasons for thinking that the job can be done quite quickly.

In a 1982 petition published in the journal *Science* the late cosmologist Carl Sagan wrote: 'We are unanimous in our conviction that the only significant test of the existence of extraterrestrial intelligence is an experimental one. No a priori arguments on this subject can be compelling or should be used as a substitute for an observational program.' That was a petition in support of the SETI program, but it has equal force in the search for life on Mars, and in the quest for the origin of life on Earth. Argument and speculation can be fun around the dinner table, but to make progress we have to look.

The discovery of evidence of former life on Mars could be interpreted in several ways: the first is that life arose on Mars independently of that on Earth. That would be every bit as significant as the discovery of life in the more distant realms of the Universe. A second possibility, in some ways even more startling, would be that early life on Mars could be found to be indistinguishable from that on Earth. That could re-ignite interest in the Panspermia hypothesis which suggests that the Universe is pregnant with life and it moves from one heavenly body to another aboard comets and other objects. More likely it would focus attention on the possible transfer of life between the planets aboard meteorites. It would also give a great boost to research on self-organising systems. These are systems or processes in which the laws of physics and chemistry when coupled with the effects of Darwinian selection might drive evolution in particular directions, resulting in familiar outcomes. I return to these issues in chapter 8.

We are at an early stage in our exploration for earliest life on Earth, and of course the search for life on Mars has barely begun. So there are a great many uncertainties. This is not a field for the timid, the pessimistic or the cynical. While scepticism is an essential part of science, the advances that have been made have been won as a result of optimism, vision and determination (tinged with ambition and other more venal qualities). The great space exploration programs of the United States and the former Soviet Union, and more lately other nations, are sustained by complex motivations. But at their heart is the human quest for knowledge and understanding. Mankind as a whole is embarking on a great program of exploration of the Solar System, and the search for life elsewhere is a central driving force in that program.

The prospects are immediate: three missions to Mars were launched in 1996, two American and one Russian. The Russian vehicle failed and crashed into the Pacific, but hopefully their exploration program will survive. The United States vehicles reached Mars in July and September

1997. A Japanese mission was launched in July 1998. Many more launches are planned for the near future, by the United States, Russia, Japan and the European Space Agency, as I discuss in chapter 7. Late in 1995 the Administrator of NASA (National Aeronautics and Space Administration of the United States) announced that a 'sample-return' mission to Mars will be launched in 2005. Missions which return rocks to Earth almost certainly will be required to determine whether there once was life on Mars. There seems to be momentum building for 'crewed' missions to Mars. It is being suggested that this could happen as early as 2011.

WHY FOCUS ON MARS?

Mars is close enough to Earth that we can reasonably expect in the near future to be able to send scientists there to undertake intensive studies such as might be required in the search for life. But the main reason for focusing on this planet is that we have learned that at least early in its history it had a climate potentially amenable to life. Mars now is a frigid desert. But it was not always so.

In order to understand the search for life on Mars we need to start with a sketch of the planet. Mars is the fourth planet from the Sun, with an elliptical orbit ranging from 207 to 249 million km from the Sun. Earth is much closer (147 to 152 million km) and therefore warmer. It takes 687 Earth days for Mars to orbit the Sun (670 Martian days, or 'sols'). A day on Mars lasts 24 hours and 37 minutes. The diameter of the planet is 6780 km, about half that of Earth. Its surface area is about the same as that of the land area on Earth. At present its axis is inclined at 25 degrees to the ecliptic (the plane of rotation around the Sun), much like Earth's, and so it has similar seasons. Southern hemisphere springs and summers are shorter but much hotter than those of the northern hemisphere, with peak temperatures as much as 30°C higher, because they occur when the planet is closer to the Sun.

At low latitudes now the daily temperatures range from about −100°C to +17°C, and the average is −60°C. Because of the low pressure exerted by the very thin atmosphere, at these temperatures liquid water is unstable. Consequently, the water ice at the poles sublimes (goes straight from ice to vapour) into the atmosphere. Down to latitudes of about 40°, ice can exist in the ground as 'permafrost' as shallow as one metre. Water ice has been detected at the north pole when it is exposed as the overlying carbon dioxide ice ('dry ice') sublimes in summer.

The atmosphere is 95 per cent carbon dioxide, 2.7 per cent nitrogen, 1.6 per cent argon, 0.13 per cent oxygen and contains minute traces of other gases. In contrast, Earth's atmosphere is 78.1 per cent nitrogen, 20.9 per cent oxygen, 0.93 per cent argon, and 0.03 per cent carbon dioxide. That of Mars is thin, with the pressure at the surface of the planet (a mean of 5.6 millibars) being less than one hundredth of that on Earth. So, if we landed on the surface and our pressure suits failed our blood would boil, the carbon dioxide would poison us, and we would asphyxiate from lack of oxygen. There is insufficient ozone to form a shield to block lethal levels of ultraviolet radiation from the Sun, there is very little magnetic field so no 'magnetosphere' to divert cancer-inducing cosmic radiation, and the red surface of the planet is highly oxidising and would destroy any organic compounds. The surface of Mars is not a great place to live. Underground, though, there are likely to be suitable habitats for microbial life, just as there are on Earth (chapter 2).

It has been calculated that over the last 10 million years the angle of the spin axis of Mars (the 'obliquity') to the ecliptic has ranged from 13° to 47°. The obliquity varies chaotically, on a time scale of hundreds of thousands to millions of years. In contrast, the obliquity of Earth is stabilised by the presence of the Moon. When the obliquity of Mars is at a minimum the poles would have permanent caps of frozen carbon dioxide, because as on Earth little of the Sun's warmth would reach the poles; when it is at a maximum, the polar caps would melt in summer. At times of high obliquity the water and carbon dioxide stored at the poles would vaporise and be released into the atmosphere, possibly raising the pressure high enough to make liquid water stable for short times. At such times any subsurface microbiota that might exist could migrate to the surface (chapter 6). So even in recent times there could have been habitable places on the surface, such as lakes and springs. Even now, landslides on the sides of volcanoes and in canyons could, for a brief time, allow the exposure of subsurface aquifers and any microbes they might contain. It is worth remembering that in one of the driest deserts on Earth, the Namib in southern Africa, life is sustained by water mists which drift over the desert from the Atlantic Ocean.

In 1976 *Viking*s *1* and *2* searched for life, but found none (chapter 7). But they and other missions found powerful evidence that early in its history, Mars was a warmer and wetter place, and therefore potentially habitable. It is necessary first to explain how we can know the age of features on Mars. All the inner, rocky planets in the Solar System were formed by the infall of rocky debris, from the size of dust grains up to that of small planets. We can see one result of this on the Moon: lots of impact craters. The number of impacts was very high as the planets were forming, and then decreased sharply once most of the rocky debris in the Solar System had been swept up. A result of the process is that old parts of a planet or moon are heavily cratered, whereas younger parts have fewer and fewer craters. Most of this record has been obliterated on Earth by later geological processes, but on a dead place like the Moon we can see the record preserved; nothing much has happened there for the last few billion years. Different parts of the Moon have different abundances of craters. Rocks from these areas were collected by the *Apollo* astronauts and dated back here on Earth. So now we can relate crater abundance to age. This provides us with a crude clock that we can use on other planets, if we make the assumption that cratering history was uniform at least across the inner parts of the Solar System. By counting the number of craters on a terrain, the age of its surface can be roughly estimated.

In some ways Mars is like the Moon. Its internal heat, relict from the kinetic energy of planetary accretion and also generated by the decay of radioactive elements, dissipated rapidly. On Earth that internal heat causes the inner parts of the planet to flow convectively, driving tectonic, mountain-building processes. On Mars those processes never developed. The crust of Mars is not constantly reworked, in contrast to that of Earth. So two-thirds of the planet has landscapes more than 3500 million years old. On Earth, rocks that old are rare, most have been greatly altered by later heating and folding within the crust, and landscapes can only be imagined, not observed.

Martian landscapes can be dated to three periods of its history, called Noachian, Hesperian and Amazonian. The absolute ages of these periods are not accurately known, and estimates depend on what assumptions are made about cratering rates. The Noachian is the oldest, and is older than 3500 to 4300 million years. This is

9

followed by the Hesperian, older than 1800 to 3500, and the Amazonian is the youngest. Noachian landscapes form the highlands of the southern hemisphere, whereas the lowlands of the northern hemisphere are younger. The oldest surfaces have abundant evidence of the former presence of liquid water. First, the craters on these surfaces are deeply eroded, in contrast to those on younger surfaces that are very well preserved. The climate must have been different from now to cause the erosion, probably by rain and wind. Second, there are numerous networks of dry valleys. These are comparable to river valleys on Earth, and though their origin is debated it is generally considered that they were eroded by rivers (of water). An alternative is that they might have been formed by 'mass wasting', the flow of water-logged sediment. However, the recent discovery of a channel within at least one of these valleys strongly supports the river-erosion interpretation. (See Figure 1.1.)

There are also giant river courses considered to have resulted from brief, catastrophic floods, perhaps when a meteorite impact melted large amounts of permafrost, or underground ice, or sent a shock wave through an aquifer. These are called outflow channels, and are up to tens of kilometres wide. They start wide and branch downstream, and where they flow over level plains spread out to be hundreds of kilometres across. Many start in 'chaotic terrain' which looks as if the ground has collapsed. They are comparable to some giant channels on Earth that formed when natural dams collapsed and lakes drained catastrophically. But some of the floods were 100 times larger than the largest known terrestrial example. (See Figure 1.2.)

There are some young valleys, on steep slopes of crater and canyon walls, and on the flanks of volcanoes. There are enormous volcanoes, including Olympus Mons, at 27 km high the biggest in the Solar System

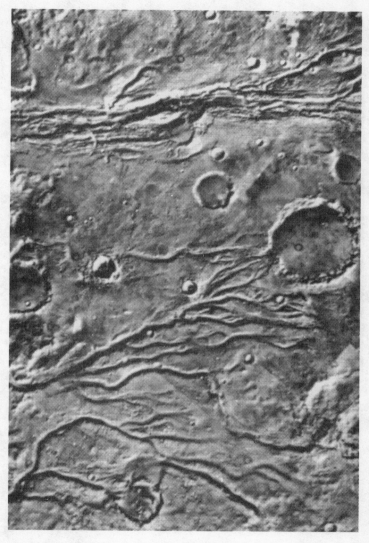

Figure 1.1 River valleys and meteorite impact craters on Mars. *Viking* mission photomosaic, JPL/NASA.

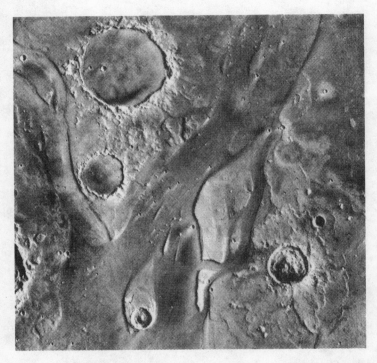

Figure 1.2 The giant 'outflow channel' Ares Vallis, a result of catastrophic flooding. The largest meteorite crater is 62 km wide. *Viking* mission photograph, NASA.

Figure 1.3 Olympus Mons, the largest volcano in the Solar System, 600 km wide and 27 km high. *Viking* mission photograph, NASA.

(and more than three times as high as Mt Everest). Though the volcanoes are currently inactive they have been active through much of Martian history. The young valleys have been used as evidence of intermittent warm and wet periods, but they might only indicate local flow from groundwater springs in areas of higher heat flow, such as near volcanoes. The water cycle on Mars is not understood, and there are strongly contrasting interpretations. The presence of ancient river channels suggests that Mars was warm and wet early in its history, but this is hard to explain. The problem is that the Sun would have been less luminous at the time the channels formed, more than 3000 million years ago (that is predicted by what we know of the evolution of stars). Even a very powerful greenhouse effect caused by abundant carbon dioxide in the atmosphere might not have been enough to warm the planet above the freezing point of water. While many scientists think the evidence indicates that Mars was warm and wet for an extended period, others suggest that such conditions might have happened only briefly and intermittently throughout Martian history. It is argued by some scientists that from time to time giant floods caused by volcanic heating of permafrost flooded the lowlands of the northern hemisphere and produced a temporary ocean.

So there is abundant evidence for the former presence of liquid water, and water ice has been directly observed at the north pole. But estimates of the quantity of water on Mars vary widely. The leading expert on this subject is Michael Carr of the US Geological Survey. He quotes estimates that range from enough water to cover the whole of Mars to depths of more than 10 km, all the way down to less than 10 m. The erosional features that we observe seem to require the equivalent of 'a few hundred metres' of water. This is an estimate of the total quantity of water. It is not suggested that it was evenly distributed across the planet. It could not have been, because the channels and other erosional features formed on land. Where is this water now? Some was lost by reacting with rocks to form clays and other water-bearing minerals. Some would have been broken down and literally blown away from the upper atmosphere by the solar wind. Some is present as ice at the poles. But most is probably underground, as ice. At high latitudes this permafrost could be hundreds of metres thick. Impacts on Mars during the late stage of planetary accretion would have fractured much of the crust to depths of kilometres, creating a highly porous and permeable veneer of broken rock, a 'regolith' into which the water would have soaked.

The magnetic field of a planet tells us much about the composition and evolution of the planet, and its habitability. Until recently it was thought that Mars has little or no magnetic field. Magnetisation detected in some meteorites considered to have come from Mars (chapter 5) hinted at the possibility that there might have been a stronger field in the past. Measurements by the *Global Surveyor* orbiter early in 1998 have now revealed that there are large terrains on Mars that are magnetised with a field strength up to 400 nanoteslas, $\frac{1}{75}$ that on Earth. The terrains are a few hundred square kilometres in area, and each has a different direction of magnetisation showing that these are fields remnant from earlier times and which subsequently have been disrupted. *Mars Global Surveyor* found that the magnetic field on Mars is stronger than had been expected. But it is not being generated at present, it seems. Jack Connery of the mission team says Mars is 'strewn with multiple magnetic anomalies'. These also occur on Earth and result from past magnetisation events.

EARTH

Earth provides our model for early life on Mars, and our laboratory for testing search techniques. On Earth the oldest convincing evidence of life is found in the Pilbara district of Western Australia, and is about 3.5 billion years old (chapter 3). It includes fossilised thread-like and globular bacteria that lived in felt-like mats on the floors of lakes or lagoons. There is also evidence of bacteria at the same time living around ancient hot springs on the sea-floor. Similar evidence is found in South Africa, in rocks of much the same age. There is tentative evidence of life back to 3800 million years ago, in Greenland. That evidence is in the form of patterns of carbon isotopes. Very little is known about life from these times because we know of few well-preserved rocks that contain the record, and there has been little research on those rocks that are known.

Many biologists now predict that the earliest life on Earth lived at high temperatures, and that life might have originated in hot springs (chapter 2). Palaeontologists have yet to study ancient examples of such environments. Nonetheless, we do know that until about 1000 million years ago almost all life on Earth was microscopic. So from whenever life originated—perhaps 4000 million years ago—to 1000 million years ago, no animal eye was present to view the scene, no kelp forests or corals filled the seas, no tree grew on land. To our eyes Earth would have seemed barren. Yet it teemed with life. Wherever there was liquid water cooler than about 150°C, there were microbes. And that is what I expect we will find on Mars.

So this is a book about finding microscopic life; the search is about looking for needles in haystacks. You might think that it is an immense arrogance to think that it is possible to find a fossil bacterium on Mars, but it is, because palaeontologists have done it here many times. You might also think that it does not matter much. I hope to convince you otherwise. If you despair of finding anything of interest in a book about the search for microbes, skip straight to chapter 8 and the Epilogue. That might change your mind. I hope so.

This is not a book about the 'pyramids' and the 'face' on Mars, familiar to many enthusiastic believers in a populated Universe; these are just natural landforms. One more bit of housekeeping: I'm not part of an international conspiracy to cover up the discovery of 'aliens', an accusation occasionally levelled at scientists. If I thought aliens existed on Mars, I would be delighted to tell you about them.

What follows is first a discussion of what is known about early life on Earth, from the predictions of biology and the geological record, then the evidence from Martian meteorites, the search strategy for Mars, mission plans, and finally some thoughts on the significance of it all. At the end is a list of a few of the interesting books I have read on these subjects, and a list of websites and electronic newsletters.

The exploration of the Solar System, and the search for life beyond Earth, are two of the greatest ventures in human history, but it will be a long time before we have a dispassionate perspective. Meanwhile, it's a great story, and here is my version of one part of it.

CHAPTER 2

THE UNIVERSAL TREE OF LIFE

Discovering the steps in the origin and earliest evolution of life on Earth might be possible if we had a complete historical record in ancient rocks. Unfortunately, as I discuss in the next chapter, we have almost no record at all. Hardly any rocks survive from the appropriate times. It is like a history book with most of the pages missing, so we have to supplement the meagre fossil record somehow.

In 1965 E. Zuckerkandl and Linus Pauling published an article entitled 'Molecules as documents of evolutionary history'. Genetic molecules contain the blueprint for cells, the hereditary code. Small changes in these molecules are the source of biological diversity. Environmental pressures act on this diversity to bring about evolution. As a result, comparisons of the compositions of genes or the proteins they code for can be used to infer evolutionary relationships. By this method, Zuckerkandl and Pauling suggested, the genealogy of life could be revealed.

Constructing such a genealogy, or phylogeny as it is called technically, is also one of the goals of palaeontology. Palaeontologists have been very successful in uncovering the history of those organisms that had bits that fossilised: skeletons, shells, bones, leaves, wood and so on—even flowers, pollen and feathers. But when it comes to the history of microbes with no hard parts, so far we have found only a sketchy record. It turns out, as I will discuss later, that microbes make up much of life on Earth, and were the sum total of life for the first two billion years or more of Earth's biological history. So while palaeontologists are making great strides in this area, at present it falls to biologists to discover most aspects of the phylogeny of microbial life.

The obvious molecule to analyse and compare in order to establish a phylogeny is DNA, the molecule that stores the genetic information in all forms of life (except some viruses). The problem with this is that DNA molecules are huge and analysing them is a monumental job. The Human Genome Project is doing just that at present for our own DNA, and the work has occupied major laboratories around the world for years and is still not complete. Microbes have less elaborate DNA than humans but even so the analytical task is daunting. Only a handful of species of microbes have been completely 'sequenced' (where the order of the 'bases' making up the code is determined) so far. To give some idea of the magnitude of the task, the article describing the DNA of the microbe *Methanococcus jannaschii* has 40 authors, in contrast to the usual one or two!

What is needed is an information-rich molecule that occurs in all organisms but is small enough that analysing it is not a lifetime's work. There are a number of possibilities, and a range of different molecules have been utilised. The analysis of one in particular has revolutionised biology. It is called a small sub-unit of ribosomal RNA (16S rRNA). Ribosomes are particles that occur in all cells, and are the site of protein synthesis, that is, where proteins are made. The genetic molecule RNA contains the code needed to construct the different proteins required by the cell. That code, and the proteins, differ between different species, and

19

the degree of difference is a measure of the relatedness of the organisms. The logic is that ribosomes occur in all cells and therefore are a primitive characteristic. The differences between the ribosomes of different species result from the evolutionary processes that produce new species. A hundred or more species, representative of all known major forms of life, have now been analysed, allowing the mapping of the 'Universal Tree of Life' (see Figure 2.1).

The pioneer of this method, Carl Woese of the University of Illinois, started his laborious task in 1966, but it is only in the last five years that the enormousness of his achievement has been widely recognised. As the German chemist and researcher of the origin of life Günter Wächtershäuser has said, 'It's as if Woese lifted a whole submerged continent out of the ocean'. Now, for the first time, we have a comprehensive chart of the evolutionary relationships of all the main kinds of living organisms. The techniques available to do this in the 1960s were very labour-intensive, and have been replaced by much faster ones in which the ribosomal RNA genes in DNA are compared. This research is now done in many laboratories, but it took nearly 30 years of determination by Woese to lay the groundwork. There are still many problems with this approach. For example, genes can be transferred between species, scrambling the record of evolution (see chapter 4). Nonetheless, the basic pattern of evolution seems to have been revealed.

So, what is life? It turns out that, overwhelmingly, it is microbes. Currently, three great superkingdoms, or domains, of life are recognised, the Bacteria, Archaea, and Eucarya. The twigs on the tree of life represent the organisms that have been sequenced, and most of them are microbes. Animals and plants lie at the end of one great branch, Eucarya, represented by us (*Homo*), and maize (*Zea*). On this scale our closest relatives are mush-

Figure 2.1 'Universal tree of life'. Comparison of ribosomal DNA of 64 species representative of the three superkingdoms of life allowed the construction of this chart of the relationships of different organisms. The longer the branches, the greater the differences.

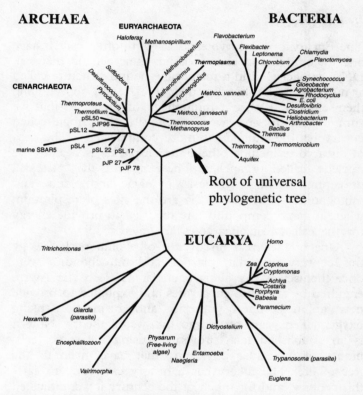

Source: Adapted from Barnes et al. (1996) in *Evolution of Hydrothermal Ecosystems on Earth and Mars*, John Wiley & Sons

rooms (*Coprinus*). If all the other large organisms we normally think of when we use the word 'life' were plotted on the tree, they would all be crowded amongst and around these three twigs. While we take a lot of interest in such 'macroscopic' life, for a host of very good reasons, in a sense we have until recently missed the point. Most life is microscopic, even single-celled.

EUCARYOTES

The Eucarya, or eucaryotes as they are often called, have one major defining characteristic: their genetic material, DNA, is packaged in a membrane-bound structure called a nucleus. In replicating their DNA and reproducing themselves, most use a multiplicity of processes that we sum up with one word: sex (in the single-celled eucaryotes the situation is more complicated). The cells of eucaryotes include other internal packages besides the nucleus, called organelles. These include the 'plastids', or centres of photosynthesis, of algae and plants, and 'mitochondria'. Mitochondria are the sites of respiration, where organic compounds are oxidised to provide energy for the cell. Most eucaryotes are microscopic.

One of the great discoveries of evolutionary biology this century was that plastids and mitochondria were once themselves free-living bacteria. These bacteria were engulfed by ancestral eucaryotes and exploited to provide new abilities: photosynthesis, and respiration using oxygen. Ten years or so ago this was a theory. Now it is an established fact, proven by using the same techniques of molecular phylogeny that have given us the tree of life. The theory has a history extending to early this century, and for much of the century it was ridiculed by most biologists. One who did not ridicule it was the American biologist Lynn Margulis. She became its champion, from the late 1960s, and still is. The process is called 'endosymbiotic evolution'. To my shame, when I was a student at the University of Adelaide in about 1970, my professor Martin Glaessner gave me this issue as a seminar topic, and I failed to see what all the fuss was about.

The engulfing and incorporation of plastids and mitochondria, and possibly other organelles, in a number of evolutionary lineages and possibly on a number of occasions, points to one of the problems in reconstructing the tree of life. Evolution was not a strictly linear process. There was some crisscrossing between lineages. In the example of organelles this web of interrelationships is being sorted out using both the comparison of the structures of whole microbes and organelles and the techniques of molecular phylogeny. But there is a more difficult problem: cells of different species, even different superkingdoms, can exchange genetic material. This is called lateral gene transfer, and I discuss it in a later chapter. But the result is, at least to some extent, that rather than a tree of life, we have a web of life. The prevailing view amongst biologists is that this does not negate the interpretations of Carl Woese and others, but it adds a substantial complication, especially to our attempts to unravel the earliest history of life when there may have been fewer constraints on lateral gene transfer.

BACTERIA AND ARCHAEA

Bacteria and Archaea used to be lumped together as procaryotes, and some biologists still do so. These organisms contrast with eucaryotes in lacking nuclei, other organelles, and sexual reproduction. They are all microscopic, with very rare and then only marginal exceptions. Until recently they were all called bacteria, and often still are. But their superficial similarities conceal profound differences.

Archaeons have more in common with eucaryotes than with bacteria, suggesting that the ancestral cell population of the eucaryotes was an archaeon. It was

23

this population that organised its DNA into a nucleus and engulfed bacteria to establish mitochondria and plastids. Even so, archaeons differ greatly from eucaryotes as well. The first archaeon to be completely DNA-sequenced revealed that 56 per cent of its 1738 genes were previously unknown, and different from any found in the other two superkingdoms. This is a huge difference, and vindicates Woese's separation of them from the bacteria and eucaryotes.

Amongst the deepest differences between archaeons and bacteria are the processes of reproduction and coding for proteins. The genes that control these processes are different in the two superkingdoms. On the other hand, the metabolic processes in the two groups are much the same, perhaps reflecting lateral gene transfer early in the evolution of the superkingdoms. A difference that is important to palaeontologists is the composition of the cell walls of the two groups: the bacteria and eucaryotes have much in common, but the walls of archaeons are composed of 'ether-linked lipids', a class of compounds unknown in the other groups. These differences survive decomposition of the cells and even billions of years of preservation in rocks, and have been found in rocks 1700 million years old, and maybe much older. This and other chemical compounds provide markers which are being used to uncover the history of all three superkingdoms, as we shall see in the next chapter.

In a sense I was a witness to the discovery of the Archaea. In 1971 and 1972 I was a postdoctoral student at Yale University in the United States. I was working on the relationships between the microbes of hot springs and the mineral deposits that occur in them. My field area was Yellowstone National Park, surely one of the most beautiful places on Earth. I was collaborating with microbiologist Tom Brock, then of Indiana University, and his Australian graduate student John Bauld. Until just before that time, the hottest waters in the springs were considered to be sterile. Then Brock and his students discovered a microbe they later called *Sulfolobus*, living at up to 85°C. A cascade of subsequent events led to the study of numerous microbes that thrive in what from our perspective are extreme environments: concentrated salt solutions, strong acids and alkalis, and at high temperatures. Many of these microbes had been known before, but there was a growing recognition that they had things in common. Then along came the results of Carl Woese to make sense of it all. *Sulfolobus* is an archaeon. There are aspects of this story that illustrate the obverse of scientific innovation: conservatism and rigidity. The manuscript describing *Sulfolobus* was rejected twice before it was published, because the results seemed anomalous and implausible. The editors of scientific journals have a tough job. They tread the line between rigour and recognition of innovation, and become infamous when they get it wrong.

LAST COMMON ANCESTOR

Finding the root of the tree of life, and therefore the 'last common ancestor' of all current life on Earth, presents a fundamental problem. The basal point of any particular part of the tree (animals for instance) is found by comparing the gene compositions of the branches in that part with that of an 'outgroup' (say plants). An outgroup is a related set of organisms in a different part of the tree. The point of closest similarity is the root, or base, of that area of the tree. But there is no outgroup for the whole tree. We know of no organisms with which to contrast all known life on Earth. A solution to this problem was devised simultaneously by Naoyuki Iwabe and colleagues and Peter Gogartan and colleagues in 1989. Although there are no outgroup organisms, there are outgroup genes. The idea is that if some gene in the population of organisms ancestral to all life duplicated itself, and the duplicates evolved into different forms with the passage of time, then those two genetic lineages could

Figure 2.2 Simplified version of the universal tree of life showing the major groups. The plastids and mitochondria of the Eucarya arose by the incorporation of bacteria. All the most ancient branches are occupied by hyperthermophiles. Therefore, the last common ancestor (cenancestor) was also a hyperthermophile.

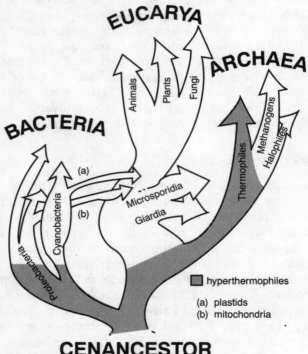

Source: Adapted from Doolittle and Brown (1997) in *The Origin and Early Evolution of Life*, Pontifical Academy of Science.

be compared to find the point of divergence, or root of all life. Iwabe, Gogartan and colleagues, and others subsequently, identified three such genes, and by this method the base of the tree of life was found (see Figure 2.2). However, biologists are at an early stage in this research and there are strong arguments against the interpretation just presented. We will have to wait to see whether more research vindicates this view. In chapter 4 I describe one view of what the last common ancestor might have been like.

It is reasonable to infer that characteristics shared by organisms closest to the lowest branches of the tree of life occurred in the community of organisms represented by that branch point, the last common ancestor. Deep lineages of eucaryotes have not yet been discovered. Analysis of the genes of bacteria and archaeons reveals that these two superkingdoms share basic mechanisms for growth, or 'metabolism'. Of those organisms nearest the base of the tree of life (as presently recognised) that have been grown in culture in the laboratory, all can use hydrogen to reduce carbon dioxide to manufacture their cell material (a form of the process called 'autotrophy'). All the lowest branches of the tree are occupied by hyperthermophiles, a term introduced by one of the pioneers in the field, German microbiologist Karl Stetter. These organisms grow best at temperatures of more than 80°C. So the last common ancestor was probably a heat-loving autotroph. The upper temperature limit of life is not known, but it is at least 113°C, and may be 150° or more.

MICROBIAL WORLD

It is sobering to consider how little we know about the microbial world. Whole ecosystems are still being discovered. That of the deep sea black smokers was found

27

only twenty years ago. And now in the last five years research has started to reveal that rocks as deep as 2–3 km beneath Earth's surface are populated by microbes (there have been hints of this for a long time, but only recently have new techniques allowed convincing interpretations). The use of new 'gene probe' techniques is showing that only a tiny fraction of microbial life has been described. American microbiologist Norman Pace estimates that even in 'well-known' ecosystems more than 99 per cent of microbes seen microscopically have not yet been grown in culture and so are little known. No doubt there are many surprises in store for us.

So we now have a very powerful predictive hypothesis that indicates which forms of life we can expect to find as fossils in the oldest rocks on Earth. However, there is something missing: what the tree of life does not do is indicate the timing of evolutionary events. That is because the genetic changes that enable evolution do not occur in a clock-work manner, and biologists cannot calibrate the tree against time. The rate of evolution varied greatly in the different branches. Thus to provide a time-frame we must return to the geological record.

CHAPTER 3

THE FIRST BILLION YEARS OF LIFE ON EARTH

In the beginning, as far as the Solar System is concerned, there was just a cloud of gas and dust. Under the influence of gravity, the cloud separated into clumps. The huge central clump collapsed in on itself and became so dense that nuclear reactions began. The Sun was born. Heat from the Sun drove the volatile gases to the outer reaches of the Solar System, eventually to become the gaseous planets, and comets.

Earth, Mars and the other inner planets of the Solar System formed by the agglomeration of the less volatile rocky debris, including 'planetesimals', that is, giant meteorites.

Isotopic evidence indicates that Earth achieved almost its present size during its first 50 million years, from 4550 to 4500 million years ago. The same would have been true of Mars. To achieve that, the rate of infall of meteorites, the 'impact rate', must have been a billion times that of the present. One of those impactors into Earth was as big as Mars, and some of the molten rock it blasted into space condensed to form the Moon. Subsequent tectonic processes and erosion on Earth have almost completely obliterated the record of accretion of Earth, but we can study the Moon to learn about these times. Studies of the Moon have shown that the meteorite impact rate there 4000 million years ago was hundreds of times the present rate. It would have been much the same on Earth. Accretion of Earth continued until 3800 million years ago, and until at least 4000 million years ago the incoming flux of giant meteorites would have made the surface of the growing planet uninhabitable. Some of the impactors would have released sufficient kinetic energy to vaporise the oceans.

It has been suggested that life could have arisen and have been extinguished by the impacts of giant meteorites a number of times. It has also been suggested that only hyperthermophiles living in some relatively protected environment such as deep underground or in deep submarine hot springs might have survived a giant impact, thus accounting for the fact that hyperthermophiles lie close to the root of the universal tree of life: only they survived the cataclysm and could go on to give rise to all subsequent life. These interpretations are possible but there are no tests available at present. The alternative and simpler hypothesis, explored in the next chapter, is that life arose at high temperatures. Either way it is reasonable to assume that the 'last common ancestor' of life lived no more than about 4000 million years ago. Life as we know it may have begun at that time.

THE OLDEST ROCKS

The geological record does not help us much with deciphering the very earliest stages of life. The oldest known rocks on Earth are 4020 million years old, but these have been so altered by later processes that they are unlikely to reveal anything about life. The 3800 million year old 'gneisses' of Itsaq in Greenland provide us with our first glimpse of life, maybe. The problem is that these rocks have been heated to more than 700°C, twice, and greatly deformed as well, degrading the record of the conditions under which they originally formed. Nonetheless, because they include conglomerate and other recognisable sediments, they yield convincing evidence for original deposition from liquid water, and thus an Earth surface temperature of less than the boiling point of water.

There is a long history of attempts to find fossils in the Greenland rocks. Some of the earliest attempts to use 'biomarkers', hydrocarbons remnant from chemical components of living cells, used these rocks. I will discuss this technique later in this chapter, but amino acids found in the Itsaq gneisses turned out to be contaminants, as was discovered to be so for all other reports of very ancient amino acids. The reported 'microfossils' proved to be 'fluid inclusions' in minerals, that is, tiny globules of fluid trapped by the rock crystals as they grew. More intriguing evidence has come from the carbon isotopic composition of graphite and calcium carbonate. Manfred Schidlowski of Germany is the leading advocate of using this approach on the Itsaq rocks. Many life processes favour the light isotope of carbon, ^{12}C, over that with an extra neutron, ^{13}C. The result is the 'fractionation' of carbon isotopes, causing ^{12}C to be concentrated in living cells, leaving the surrounding environment, for instance the ocean, depleted of ^{12}C. This gives us a signal for life which can be found in remnants of cells, and in carbonate minerals precipitated from ocean water. Unfortunately, heating of rocks to high temperatures converts any original cell material to graphite and degrades this signal, making it hard to interpret. The results from Greenland could be evidence of life, as Professor Schidlowski contends, but might just as well be explained as by cometary or meteoritic inputs of carbon. A recent report that tiny specks of carbon in apatite (calcium phosphate) from Itsaq have an isotopic composition indicative of biological methane production is intriguing

31

but needs to be confirmed. There are other possible explanations. For example, synthesis of organic compounds by lightning also causes fractionation of isotopes. The latest results are a good example of cutting-edge science, but it is too early to accept them uncritically. So, there are hints that there was life 3800 million years ago, but the evidence is not yet compelling.

3.5 BILLION YEARS AGO

That brings us to 3500 million years ago, the age of the oldest well-preserved sedimentary rocks known. These are in the Pilbara region of Western Australia, and the Barberton Mountainland, South Africa. Both have yielded what is generally considered to be convincing evidence of life—indeed of complex microbial ecosystems. Once again, though, when the evidence is examined critically there are many problems. We can take Western Australia as our example. The evidence is threefold: the presence of fossils called stromatolites, of fossil cells (so-called microfossils), and patterns of distribution of isotopes of carbon and sulfur. They come from a place ironically called North Pole (Plate 1) by the early gold miners who worked there: it is remote and inhospitable (in this case because the weather gets extremely hot), and they must have thought they might as well have been at the real North Pole.

Stromatolites are by far the most abundant fossils in ancient rocks, and the source of many of the key clues to the microbiology of the early Earth. Rather than direct remants of organisms, like a fossil shell or bone, they are traces of the life activities of particular kinds of microbes. They are sediments sculpted into distinctive shapes when felt-like mats of microbes living on the sea-floor or lake floors trap mud washed onto them or precipitate minerals such as calcium carbonate, as a result of their life activities. The microbes then grow or move up through the mineral blanket to establish a new mat. Many microbes can move under their own power; I have measured cyanobacteria racing along at 16 microns (thousandths of a millimetre) per minute. The movement or growth results from the need to have access to light, for photosynthesis, or access to oxygen or other chemicals found in the overlying water. The mats have distinctive shapes generated by the various patterns of division of the microbes' cells, the various ways they aggregate when they move, the erosive effects of water currents, and other effects. As a result, encoded in the shapes is a large amount of information about the biology and ecology of the constructing microbes. It is rare for cells to be preserved in stromatolites, so interpretations rely mostly on the shapes of the sedimentary layers, and this leads to many uncertainties.

Despite being built by microbes, stromatolites can reach huge sizes and even ultimately build reefs as big as the Great Barrier Reef of Australia. Such giant reefs are seen in rocks more than 2000 million years old in Canada and South Africa. Most, though, are the size of a cabbage, or somewhat larger. Cut through they also look like a cabbage, with each layer representing a former layer of microbes. We can make fairly confident interpretations of stromatolites because they can still be observed forming in a few places. The most famous modern occurrence is in Shark Bay (Plate 2) in Western Australia (their presence there is one of the reasons Shark Bay is on the World Heritage list).

The North Pole stromatolites were discovered by John Dunlop, then a student of the University of Western Australia, and were jointly studied by him, another student of the time (1978–1980) Roger Buick, and myself. At the same time, Don Lowe of Louisiana State University discovered another occurrence of stromatolites nearby and independently submitted a manuscript describing them to the British journal *Nature*. Our two articles were published in the same issue of *Nature*. Lowe could have had his published first and scooped us, but when I heard about his work I wrote and told him that the students and I were about to submit a manuscript on another occurrence of stromatolites, and he and the editors of *Nature* graciously agreed to delay publication of his paper.

Inevitably there have been disputes about the North Pole stromatolites. They are a significant part of the evidence for earliest life on Earth, and so the evidence has been scrutinised intensely. Intense scrutiny is not the same as intense study: little has been published in the twenty years since our initial work. Don Lowe has changed his mind and now thinks there are no definite stromatolites older than 3100 million years, including those he first described. The reality here is hidden behind the words. Even Lowe thinks that the form of the thinnest sedimentary layers, the laminae, may in some examples indicate the former presence of microbes. He just contends that the large-scale shapes we see out in the bush have other origins. He now interprets those he originally described as deposits like stalactites in caves, and the one I described with John Dunlop and Roger Buick as a 'fold', distorted layers of sediment. Even if we accept his reasoning, which I do not, in the small-scale form of the laminae we still have evidence for microbes.

It is difficult to prove that most of the objects identified as stromatolites were actually built by microbes. Over the decades since the word stromatolite was coined in 1908 (as 'stromatolith' by the German geologist Kalkowsky) numerous alternative interpretations have been proposed, ranging from their being sponges, to folds in the rock or chemical deposits. Over the same period there have been many detailed studies of stromatolites and their origin has been demonstrated convincingly. Some have preserved within them the cells of the microbes responsible for their construction. Close comparisons can be made with living examples; for instance, Hans Hofmann of the University of Montreal and Steve Golubic of Boston University described 1800 million year old examples from Baffin Island in Canada that can be matched point for point with modern examples from the Persian Gulf and Shark Bay built by the cyanobacterium *Entophysalis*. So close is the similarity that they named the fossil cells they found in the ancient stromatolites *Eoentophysalis* (the dawn of *Entophysalis*).

Stromatolites have been shown to occupy the same ecological niches on the early Earth as those later filled by algae, corals and other organisms. They built similar reefs and other carbonate deposits. But starting maybe 1000 million years ago they became less and less abundant, probably because the constructing microbes are at the base of the food chain, and once animals had evolved the microbial mats were a good source of food. An alternative interpretation is that the seaweeds that evolved at about that time displaced the microbial mats, ending their dominance of shallow marine environments. Stromatolites survive now only in extreme environments inimical to animals (and most seaweeds), such as the highly saline environments of parts of Shark Bay and the hot springs of Yellowstone National Park, and Rotorua in New Zealand.

In recent years it has been recognised that some stromatolite-like mineral deposits previously considered to be entirely chemical in origin in fact are partly biological. A good example is the 'geyserite' that forms around boiling springs. These deposits are made of opal, hydrated silicon dioxide (white, not coloured like precious opal). Where the erupting water splashes around the rims of pools, layered deposits form. These closely resemble stromatolites. They were interpreted by me during the early 1970s as chemical precipitates, but new studies using more sophisticated techniques have shown recently that geyserite is colonised by very thin 'biofilms' of hyperthermophilic microbes that seem to play a role in constructing the deposits. They apparently act as templates on which the deposition of silica is promoted. Many other mineral deposits once thought to be chemical are also likely to have been influenced by biofilms.

No geologist now seriously disputes that most stromatolites are microbial. The problem is that word 'most'. It does not matter much for times in Earth history like 2800 to 600 million years ago, when stromatolites were extraordinarily abundant. We might make the occasional mistake, but in the overall scheme of things it will not change our understanding of biological history. But prior to 2800 million years ago few examples of stromatolites are known, at least partly because we know of few rocks of that age. As a result, every occurrence is invested with great significance in discovering the earliest history of life.

John Grotzinger of the Massachusetts Institute of Technology (MIT) is a geologist who has contributed greatly to our understanding of the early Earth. His work on stromatolites has shed light on their role in building continental shelves and reefs in past times. But he too is troubled by the problem of how to prove that a rock is a stromatolite. Working with mathematician Dan Rothman, also of MIT, Grotzinger analysed the shape of one stromatolite using the principles of fractal geometry, and showed that it could be explained without the intervention of microbes. One result was a headline on the cover of *Nature*, 'Stromatolites—fraudulent fossils'. I was responsible for that headline (well, the Editor was really, but I suggested the expression), and it came back to haunt me when some anonymous wag at the university where I graduated, in Adelaide, wrote to me withdrawing the award of my PhD because my thesis on stromatolites must have been fraudulent. The serious point is that Grotzinger and Rothman consider that they proved the 'null hypothesis', that their one stromatolite could be non-biological. At the same time, I know that Grotzinger considers most stromatolites to be biological. It might be difficult for non-scientists to deal with this apparent contradiction. Lawyers would have no problem. The question is not the truth, but the difficulty of proving it. And when it comes right down to it, this is what is going to happen in the search for life on Mars. We may well find stromatolite-like objects there, but how do we get rid of that unwanted appendage '-like objects'?

On Earth we can interpret stromatolites as old as 3100 million years with a good level of confidence. In the bed of the Wit M'folozi River in Natal there are convincing 3100 million year old stromatolites of two distinct types first described by Tom Mason and V. Von Brunn of Durban. Features in the accompanying sedimentary rocks show that they formed on what were once tidal flats and in adjacent tidal channels. Neither type of stromatolite contains fossil cells, but one has features suggestive of phototaxis, that is, movement of cells towards the Sun, and therefore evidence of photosynthesis. They are closely comparable to stromatolites constructed now by cyanobacteria, and have a carbon isotopic composition consistent with that interpretation. Their context in particular sedimentary rocks indicative of formation in places where we expect microbes, their very distinctive and meaningful shapes, and their isotopic composition, all come together to allow a cogent interpretation. This is a good example of the use of multiple criteria for demonstrating a biological origin.

Then from about 2800 million years ago on up through time the evidence is abundant and generally uncontroversial. By then the rock record itself becomes much more abundant. It can be seen as part of a continuum with the 'Proterozoic' record, that from 2500 to 544 million years ago, where thousands of occurrences of microfossils have been found in hundreds of geologic units, about 1000 different types of stromatolites are recorded, there is a well-established record in carbon and sulfur isotopes, and a developing record in hydrocarbon biomarkers. Even so, from 2800 to 2500 million years ago there are only about 20 documented occurrences of stromatolites and three documented occurrences of microfossils.

Back at North Pole, we are left with disputed occurrences of stromatolites. I think they are biological, as do many other stromatolite specialists, and we have good reasons for believing so. But good reasons are not the same as proof. My colleagues Kathleen Grey of the Geological Survey of Western Australia and Hans Hofmann of the University of Montreal have recently checked several previously undocumented occurrences of possible stromatolites in this region, and they consider that some of them are convincing. If there were micro-fossils within the probable stromatolites we would be in a better position, because that would provide additional evidence of biogenicity. There are some faint remnants of cells in those from North Pole, but these are not convincing either.

Microfossils are extremely rare at North Pole and nearby, and like the stromatolites, all have been or can be disputed. Many reported microfossils of this age from Africa and Australia are no longer accepted as biogenic. They have been re-interpreted as microscopic spheroidal globules of bitumen (ancient oil that has solidified), opal spheres that formed in gelatinous silica sediment, and so on. The Spanish chemist Juan García-Ruiz has developed techniques for growing tiny crystals, some of which look

Figure 3.1 Barium carbonate crystal aggregate grown in a silica gel, in the absence of any organisms. It is 50 micrometres wide. Compare this with the spiral cyanobacterium shown in Plate 2. (Photograph by Nick Welham, courtesy of Stephen Hyde)

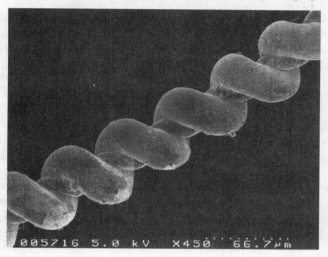

remarkably like microbes. We have to be very aware that everything that looks like a fossil is not necessarily so, a major problem we will encounter later when we discuss meteorites from Mars (chapter 5). At North Pole, filamentous forms from rocks named the Towers Formation (Plate 3) are convincingly biogenic but attempts to find more examples from the same locality have failed, so the original report has not been verified. Thus this occurrence currently fails one of the fundamental tests applied in science: reproducibility. If a result cannot be repeated by other scientists, it is regarded with great scepticism. There is also debate about whether the rocks are really the Towers Formation or whether they might be much younger intrusive 'chert dykes' emplaced when the region was folded and uplifted.

Part of the problem is relocating with certainty the exact locality from which some of the original samples were collected from the Towers Formation. In these days when geologists routinely use the Global Positioning Satellite (GPS) system to locate themselves within a few metres anywhere in the world it is difficult to appreciate how hard it used to be. We all learned the skill of accurate map-reading and the skill of interpreting aerial photographs so we could locate ourselves next to the very bush we could see on the photos. But sometimes we found ourselves in almost featureless areas and, in truth, did not really know where we were to within a few hundred metres, or even much worse. I well remember mapping the geology of the northern Simpson Desert in Australia; we had beautiful low-level colour aerial photographs on which we could see every bush. But one bush looks much like another. We figured out where we were by counting the sand dunes as we flew over them in the helicopter. I guess some day some young whippersnapper will be out there again and will discover we blinked and missed a dune. Good luck to them.

The Towers Formation samples were collected by Stan Awramik of the University of California, long before geologists used GPS navigation. Stan had had a long and productive day of collecting, and having only a few days in Australia did not want to waste any opportunities. So when a promising-looking hill of black chert loomed out of the gathering darkness at the end of the day, he did what any of us would have done, and took a sample. Murphy's Law operates in all walks of life, and so it was that this was the only sample that yielded microfossils when later studied back in Santa Barbara. Stan brought his sample down to Los Angeles to show Bill Schopf and myself, and we got very excited about it, because the microfossils are fairly well preserved and convincing. This was a major breakthrough in uncovering

40

the earliest records of life. A few months later I had an opportunity to return to Stan's locality to collect more samples. I was accompanied by Kathleen Grey, another experienced palaeontologist. We had a few problems locating the collecting site, but in the end we were confident that we had found it, at least approximately. Our samples yielded nothing. Roger Buick tried too, with the same result. So this record remains in limbo. It cannot be ignored, but it has not yet been verified.

With my colleagues from the Precambrian Paleobiology Research Group based at the University of California in Los Angeles I collected many other cherts from the Pilbara region, during 1979. None yielded any fossils as convincing as Awramik's, but there were some we regarded as probable fossil bacteria, and with Stan we published our observations in 1983. There followed more than five years of published and verbal dispute, some of which I discuss later in this chapter.

Subsequently, numerous filamentous and a smaller number of spheroidal microfossils were discovered in cherts in the Apex Basalt of the Pilbara region. Only brief descriptions of the characteristics of this occurrence, as seen in the outcrops, are available, so with this vital information missing this report too should be treated with caution. The Apex chert filaments have been attributed to a number of different taxa (biological groups) and are chains of cells 0.5 to 19.5 microns in width. They do not have the outer sheaths characteristic of many cyanobacteria and do not occur in demonstrably stromatolitic laminae. As a result, they show few features that aid interpretation. Bill Schopf, whose former student Bonnie Packer discovered the first of these, interprets the larger of them as probably cyanobacterial, because they are larger than most other forms of bacteria (and archaeons). He has made meticulous and detailed analyses of the sizes and shapes of these and many other fossil bacteria, and nobody knows more about such things than he does. However, size and shape are usually poor guides to biological affinities, particularly in the absence of other supporting evidence. In many younger examples of microfossils the shapes are sufficiently distinctive that the biological 'affinities' (relationships to well-known organisms) can be established with more confidence. But these Archaean examples are very simple—just 'balls and strings' as the aficionados say. We can push the evidence as hard as possible, like good forensic scientists in a murder case, but in the end, if there is not enough evidence we will not get a conviction.

Organic matter from both the Apex chert and the Towers Formation is depleted in ^{13}C, consistent with an origin through autotrophic metabolism. This means that there were organisms there which grew using inorganic carbon (such as carbon dioxide) as their source of carbon. The photosynthesis carried out by cyanobacteria is one such process, but not the only one, so the results do not prove that the fossils are cyanobacteria. For instance, many members of the Archaea are also autotrophs. We can be confident that there were microbes present, but what kind has yet to be proven. And at present we can make no direct link between the isotope analyses and the microfossils. In some examples they come from the same rock, but the organic matter in each rock is the end-product of a complex microbial ecosystem that probably included numerous species of diverse kinds, if the modern world is any guide. The stromatolites of the Towers Formation are likely to have been constructed by photoautotrophic filamentous microbes, based on what is known of the mechanisms of formation of extant stromatolites, but again these need not have been cyanobacteria. Even at present when few stromatolites still are forming there are examples being built by other kinds of photosynthetic bacteria.

Filamentous microfossils are also reported from 3500 million year old rocks in South Africa. Some of these occur in laminated chert that could be stromatolitic. As with those in Western Australia, the associated organic matter is depleted in ^{13}C, consistent with an autotrophic origin.

We have another source of useful evidence. In several of these ancient rock successions there are sulfur-bearing minerals. In modern environments numerous forms of bacteria and archaeons exploit the 'sulfur cycle' by oxidising and reducing sulfur compounds to gain energy for life. In so doing they fractionate the isotopes of sulfur, just as others do with carbon, as we have seen. The resultant geological record is found in sulfur-bearing sedimentary minerals such as gypsum (hydrated calcium sulfate) and pyrite (iron disulfide). Early Archaean sedimentary sulfate minerals are slightly enriched in ^{34}S as compared to associated sulfides, consistent with an origin through photosynthetic oxidation of reduced sulfur. The depletion of ^{34}S in sulfides characteristic of bacterial sulfate reduction is not seen until after 2800 million years ago.

So we can see that at North Pole and elsewhere we have a record of several kinds of microbes. Some were making cells from inorganic carbon, others were decomposing them. We have to expect that there was a complex ecosystem, but we have records of only a few components. It might seem perverse and contrary to avoid the obvious and conventional interpretation that the stromatolites are cyanobacterial and the microfossils are cyanobacteria, but there is another way to look at it: these fossils are removed by 3500 million years from living microbes we can interpret confidently, and are 'only' 500 million years or so from the origin of life. There is no continuity in the fossil record. We do not know what now-extinct microbes might have lived then, and the subsequent 1000 million years of the fossil record is very sparse, preventing us from establishing clear lineages through to convincingly interpreted fossils. We have one instant in the history of life. Almost nothing is convincingly known of earlier history, and our next good sample of life is 400 million years younger. The Pilbara fossils are extraordinarily important in our thinking about early life, but we have to be careful not to make too much of them.

The field of Precambrian palaeontology, that is the study of the fossil record of early life, is a little over 100 years old. As is the nature of science, a great many mistakes were made in the early days, and as knowledge and understanding increased, the practitioners of the field became more and more skilled, and the interpretations more cogent and durable. It is estimated by Bill Schopf that early this century more than 50 per cent of the objects reported as ancient fossils were not. The objects described are not fossils at all, or are not anywhere near as old as claimed. Sometimes this is due to the inexperience of the scientist involved, but even the old hands continue to make mistakes. Schopf estimates that even now after a century of experience some 15 per cent of reports are wrong. This is a field of palaeontology that is fundamentally different from all the rest. No-one is going to claim that the bones of a dinosaur or the petrified trunk of a tree are really not fossils at all. On the other hand, proving that a layered rock is a stromatolite or a thread-like structure is a bacterium are far more difficult problems. Isotope distribution patterns can be 'chemical' (non-biological) or biological. Even complex hydrocarbons which might be found in ancient rocks could in some cases have come from comets or meteorites rather than organisms. Sometimes the interpretation is obvious, as when fossil cells are well preserved, but very commonly several lines of evidence are required to make a compelling interpretation. This is the principle of using 'multiple criteria for biogenicity'. It has proved to be critically important for recovering the early history of life on Earth, and the same will be true of Mars. Any exploration program on Mars must be driven by this principle.

THREE BILLION YEARS OLD, AND YOUNGER

The record of Earth history younger than 3000 million years old is much richer and less controversial than the older period, because there are far more geological regions available to study. There are many stromatolites, some confidently interpreted as cyanobacterial. Microfossils are not abundant until about 2000 million years ago, but strong arguments can be made that they were cyanobacteria. There are also bizarre forms of unknown affinities. The carbon isotope record at 2500 to 2800 million years ago has been interpreted by American geochemist John Hayes to indicate the operation of oxygen-producing photosynthesis, and microbial methane production and consumption. Organic matter from several rock units of this age is highly depleted in ^{13}C. Extreme fractionation of carbon isotopes occurs during methane production by methanogenic microbes, but is not known to be produced by any other process likely to occur in natural conditions in the environments in which the rocks formed. To be incorporated into the sedimentary organic matter the methane would then have to be used as a carbon source by other microbes. That is an oxidative step that could have been coupled with the reduction of oxygen. If it involved oxygen, that would be indirect evidence of the prior existence of oxygenic photosynthesis, as no other process can produce substantial amounts of free oxygen. Such an interpretation is consistent with the stromatolite and microfossil record, and with evidence that by 2300 million years ago the atmosphere contained substantial amounts of free oxygen. That evidence is in the form of an abundance of oxidised iron in sediments, and an absence of easily oxidised minerals.

As methane production is an exclusive characteristic of some of the Archaea, we thus have evidence for the presence of that superkingdom, and all the other evidence indicates the presence of Bacteria. To date there is no published evidence for Eucarya at this time. Then for the next billion years or more cyanobacteria were widespread, building abundant stromatolites and indeed reefs.

RECONNAISSANCE

What I have written above is just the bare bones of the story of the discovery of the geological record of Archaean life. There is another side to it which is even more convoluted than the science. Careers are involved, and personal relationships. Many of the debates have certainly not been scholarly. Newspaper headlines have claimed results were 'pirated', there have been threats of prosecution for alleged illegal exports of fossils that were simply being studied somewhere other than where collected, and at times insults and rumours have been rife. Some of the damage that was done to careers and to self-esteem is probably permanent. Some disputes are still continuing.

A few useful points can be made. The first concerns the nature of reconnaissance work. There are two ways to approach a massive and difficult problem like uncovering the earliest history of life on Earth. We can try to 'pick winners' and work in great detail on a small number of sites, or we can do a wide-ranging survey which cannot include detailed studies at any sites. The work that led to many of the disputes was of the reconnaissance type, centred on the Precambrian Paleobiology Research Group led by Professor J. William Schopf. Bill Schopf assembled a group of experts from around the world and set about an ambitious, intensive but brief campaign to upgrade our understanding of the palaeobiology of the Archaean world. It was an extraordinarily multidisciplinary group, with palaeontologists, biologists, geochemists, modellers of atmospheric and oceanic chemistry, and experimenters with the origin of life. It was also multinational, with participants from the United States, Canada, Germany and Australia.

After two years of planning by a small core group, there followed 12–15 months of intense research during which some of us moved to Los Angeles and worked full-time on the project. The resultant book, *Earth's Earliest Biosphere*, and numerous associated research papers are a monument to Schopf's inspiration and leadership, and to our success, even if I do say so myself. But a lot of criticism has focused on the fact that the very extensive collection of samples on which our research was based was made without detailed studies of the collecting sites. Our mode of operation necessitated that, but where possible we worked cooperatively with local experts, and benefited from their knowledge. A few locals resented our intrusion onto their patches, and saw us as lavishly funded American big science scooping the cream off the issues that they had been meticulously studying for years. Some of my own countrymen took this attitude. Among the many lessons I learnt were these: reconnaissance does work, but it can be seen as trespassing.

What shifted the scientific debates about the Australian evidence away from any semblance of objectivity and into raw emotion was the huge interest in our discoveries shown by the media. We brought this on ourselves, by agreeing to a press release by the university's public relations group, but we certainly did not anticipate the consequences. During the time we were together in Los Angeles we did probably a hundred radio interviews, dozens of interviews for the print media, and several for TV news and documentaries. We were on the front pages of the *Los Angeles Times*, the *New York Times*, the *Times* of London, *Le Monde* and the Bangkok *Times*, to mention just a few. We were constantly hyped up. We tried to credit our colleagues outside the group, but occasionally forgot. We also learned that journalists and editors like heroes, not lists of collaborators. So more than once all hell broke loose when someone outside the group felt slighted. Some of those incidents dog me to this day, nineteen years later. Jealousy, rivalry and envy came to the surface. Some of what might have appeared objective was thinly veiled rancour. We are seeing it again with Martian meteorite ALH84001, though in that case the protagonists have stooped to even lower levels of abuse, as for instance in a cover story in *Newsweek* magazine, where insults were traded (chapter 5). And just imagine what could happen with any putative discovery of fossils on Mars.

TOP DOWN

Andrew Knoll of Harvard University is the prime advocate of an alternative to reconnaissance, which he calls 'top down'. That means working progressively further back in time from what we know reasonably well, and studying relatively small areas in detail. It is an approach that has worked very well for him and has produced elegant and cogent interpretations of life and the environment of Earth just before animals made their first mark, a billion to about 550 million years ago. He is continuing to work back in time. Early in his career he worked on much older rocks, but like the rest of us had

48

trouble coming up with convincing results. He is currently working (with me) on rocks 1700 million years old in northern Australia, so he has another two billion years of history to deal with before he works his way back to near the origin of life. Meanwhile, I don't have his patience.

My confidence in the interpretation of the very incomplete record of life in Archaean times (3800–2500 million years ago) comes from three main sources: combining multiple lines of evidence to construct internally consistent interpretations, working back (as far as possible) from the younger record which is abundant and well documented, and utilising a century of experience.

SAMPLING THE ENVIRONMENT

Palaeontology is like oil exploration, a very conservative business. People look for fossils, and oil, in the sort of places they have been found in before. This leads to diminishing returns as we run out of familiar places in which to look. All described stromatolites and microfossils and most of the isotopic data from the Archaean are from what were shallow marine and lake environments. With few possible exceptions, all represent samples of organisms that lived at 'normal' temperatures. Given the recognition by biologists of the significance of high-temperature environments as the possible cradle of life, it might be thought that geologists would have searched appropriate Archaean deposits for evidence of thermophiles and hyperthermophiles. That has not happened. To put this in context, although there are some 5000 palaeontologists in the world, only about 50 claim to be working on the early history of life and most of these do not devote themselves full-time to this subject—at the most 50 people are trying to uncover the fossil record of the Bacteria and Archaea—the other 99 per cent of palaeobiologists are working on the Eucarya.

Despite the fact that palaeobiologists have paid almost no attention to high-temperature deposits, fossils have been reported from thirteen high-temperature 'hydrothermal' mineral deposits, mostly relatively young in geological terms. There are many opportunities to extend this record back in time: there are several hundred documented occurrences of hydrothermal mineral deposits of Archaean and Proterozoic age, that is, older than 544 million years. I discuss one of them in chapter 6.

Modern hot spring environments (Plate 4) are known to have diverse biological communities, despite the fact that only the surface of these systems has been studied. The third dimension, the much more voluminous subterranean system, remains to be explored. Here palaeobiologists have a special opportunity, as examples of these systems abound in the geological record, and are much more accessible than in modern, active vents. Microbial life must be expected in this environment, just as it has been found in other subterranean settings (chapter 2).

The potential for extending the record of hyperthermophilic organisms back as far as the oldest well-preserved rocks on Earth is excellent. For instance, 3260 million year old 'volcanogenic massive sulfide' deposits in Western Australia contain exceptionally well preserved mineral textures indistinguishable from those of modern black smokers on the deep sea-floor. They were formed in at least 1000 m of water. Filamentous possible microfossils occur in these deposits, but they have yet to be studied by palaeontologists.

BIOMARKERS

Some of the organic compounds which make up organisms are so stable that they can, and do, last for billions of years after the organisms die and decompose. It is not the whole molecule that survives, but the stable 'backbone' of carbon atoms. These backbones have distinctive geometries, and finding one of these molecules is just as informative as finding a dinosaur bone. Typical examples are the 'lipids' that comprise cell walls, and pigments such as chlorophyll that absorb light to power photosynthesis. Such molecules vary in both major and subtle ways between different groups of organisms, including microbes, and they make excellent fossils. They are called biomarkers, and are very common. They are part of what oil is made of, and they occur in tiny quantities in many types of rock. Because analytical techniques are so sensitive only tiny amounts are needed. The study of these is a relatively new field but already great advances have been made.

A major limitation to the use of biomarkers is that they decompose at temperatures of more than about 200°C. Most of the Archaean rocks on Earth have been heated to temperatures higher than this, so no longer contain biomarkers. Some Archaean rocks are well enough preserved though. No Archaean biomarkers are known yet, but it is just a matter of time before some are found. A pioneer in this field is Roger Summons of the Australian Geological Survey Organisation in Canberra. I well remember when he started work on very ancient biomarkers fifteen years ago he despaired of finding any in the rocks he wished to study, more than 550 million years old. Now numerous examples are known, many found by him. (In just the last few months there has been a report of Archaean biomarkers.)

The temperature limitation will not be such a serious problem on Mars, because the lack of plate tectonics there means that rocks have not been recycled down into hot parts of the crust, unlike ancient rocks on Earth.

A MICROBIAL WORLD

In summary then:

- Life existed 3500 million years ago, and may have been present several hundred million years before then;
- Microbial mats grew in shallow lagoonal environments, and probably elsewhere.
- The microbes were both globular and filamentous.
- Some of the microbes may have used light as a source of energy, though this is more contentious; carbon isotopic evidence suggests that they were 'autotrophic', probably photosynthetic.
- By 2800 million years ago both Bacteria and Archaea were present, but so far there is no sign of Eucarya. (During early 1999 biomarker evidence of Eucarya this old was reported at a conference.)
- For at least 3 billion years life on Earth was dominated by microbes. (Plate 5)

This is our model for the search for former life on Mars. We know what types of rocks to search for, and the techniques to apply. There is also a great deal of experience with mistaken interpretations. In this immature field of study, mistakes were inevitable, but more work at the collecting sites would have prevented some of them. Particular traps laid by Nature include the presence of non-biological mineral structures that look like microfossils and stromatolites, and natural contamination of ancient rocks by younger materials such as amino acids. Probably the biggest trap though is our own egos.

To find a beautiful fossil, and then hitch your reputation to it, is no longer to see the fossil.

Palaeontologist Bob Brain, quoted by Bruce Chatwin in *The Songlines*, Picador, 1988

CHAPTER 4
THE ORIGIN

This flow of creation, from where it did arise,
Whether it was ordered or was not,
He, the Observer, in the highest heaven,
He alone knows, unless . . . He knows it not.

Hymns from the Rig-Veda, translated from the Sanskrit
by Jean LeMée, Jonathan Cape, 1975

In book reviews we read of authors in deft control of
their material, confident in their knowledge, assured in
their style. Well, maybe one of them should have written
this chapter. The origin of life is a tough subject. The
reason for tackling it here is to allow us to assess whether
it might have happened on Mars.

Consider these barriers to knowledge: I cannot see
the slightest chance that the process of biopoesis, getting
life from non-life, will ever be found recorded in rocks
on Earth, or reproduced experimentally. Knowledge from
a range of scientific disciplines can be assembled to
establish hypotheses of the origin of life, but there are
almost as many hypotheses as hypothesisers. History is
contingent, a potent mix of a vast number of variables
and insufficient knowledge. Only one hypothesis will be
correct for extant life on Earth, and there may have been
other paths to life in other places.

Biologists have given us the grand design of the universal tree of life, and can conjecture about the nature of the last common ancestor. But that is still a long way from the origin of life, and may well be a barrier beyond which they cannot penetrate, except with inspired speculation. Geologists are limited by a rock record that barely exists before 3500 million years ago, when life was already well established. Little of what happened before is recorded. Chemists can produce some of the complex organic compounds found in cells, but while this may demonstrate the feasibility of a particular reaction it does not prove that it is what actually happened in Nature. Theorists can examine self-organising complex systems and perhaps mimic some of the possible pathways to life, but showing that these were the pathways actually followed may be impossible. There is a huge number of variables and uncertainties—we know so little about the earliest Earth. We are unlikely ever to have detailed knowledge about the surface environment of Earth at the time of the origin of life.

This is a rather pessimistic view, and there is no doubt that a great deal has been learned over the last century since Charles Darwin wrote in 1871 to his friend Joseph Hooker of that 'warm little pond' of primaeval broth wherein life might have started. Much more will be learned, but it seems to me the best we can do is to provide constraints to limit the conjecture. And maybe by demonstrating our ability to make complex organic molecules, or membranes, comparable in some ways to those in cells, we can at least show that the very notion of evolving from chemicals to life is both reasonable and likely to have occurred. But there is so much contingency, so much chance, in life, that chance itself can be seen as one of the rules of Nature.

On the other hand, if life is an inevitable outcome of physical laws, perhaps its origin can be revealed. But

I doubt it. The barriers to knowledge may be like that in physics. In 1926 the German physicist Werner Heisenburg formulated the Uncertainty Principle. The very act of observing the world alters it, because of the effects of the light, or any other electromagnetic energy, with which the observation is made. For instance, if we try to pin down exactly the location of one of those fundamental particles of matter of which atoms are comprised, we cannot. The best we can do is predict where they are most likely to be. As a result we cannot precisely know the current state of the world, we can only predict its most probable state. The British physicist Stephen Hawking expressed it thus: 'one certainly cannot predict future events if one cannot even measure the present state of the universe precisely!' Physicists have dealt with this by developing the theory of quantum mechanics, which allows the probabilities of outcomes to be predicted precisely while barring us from knowledge of the exact pathways to those results. Much the same is probably true of the origin of life: we might determine that life is an almost inevitable product on appropriate planets with the right mix of compositions and surface conditions, but we might never know how it happened.

Nonetheless, a great deal has been learned, and more will follow. I have chosen to discuss the origin of life after outlining in earlier chapters the biological and geological evidence of early life and environments, because these provide us with powerful constraints to our conjectures.

LAST COMMON ANCESTOR

In seeking further constraints it is useful to start with predictions about the earliest parts of the tree of life. The branches of the tree all converge on one point, near the base. As discussed in chapter 2, this is the 'last common ancestor', the most recent ancestor shared by all three superkingdoms of life on Earth.

The last common ancestor was not a single cell from which all others descended. We cannot expect to find a genealogy like that in Genesis, where Adam begat Seth who begat Enosh who begat Kenan and so on until now. The ancestor was probably a loosely knit population of individual 'protocells'.

Carl Woese is credited with much of the original mapping of the tree of life. He has recently speculated about the last common ancestor. He imagines that it was a community of 'progenotes', proto-organisms, 'a genetically rich, distributed, communal ancestor'. In this community there was great genetic fluidity, with genes not only being passed on to progeny but also frequently passing laterally between individuals and populations. There was nothing that would be recognised as a genetically constrained species, nor even well-defined organisms as we understand them now. It was a promiscuous microbial orgy in which there were few rules. There were webs of interrelationships, not lineages.

Progenotes were unlike modern organisms. They were amalgams of components from multiple sources, with differing metabolic biochemistries peppered through the community. Innovations spread readily, by lateral gene transfer, conferring great adaptability and allowing rapid evolution. It was like an anarchic, wildly creative society practising free love. Fragments of genomes (the totality of genes in a cell) were shared freely, like viruses. Even potentially fatal mutations added to the genetic diversity of the community, maximising evolutionary potential. It might be compared to the World Wide Web, with data, programs and computer viruses spreading rapidly around the world. Like the Web, it evolved very rapidly into an ordered system.

The machinery for 'gene expression' (using the

information in genes to make proteins) and genome replication, reproduction, already existed, in some relatively simple form. This allowed favourable genomes to multiply and spread. But the error rates in genome replication and in gene translation to make proteins were high. These mechanisms improved, step by step. The improvements spread rapidly, by lateral transfer of genes. The new proteins provided the basis for more effective biochemistries. Evolution 'ceased to be constrained by imprecise translation, and progenotes, by definition, became genotes'. Somewhere along the way the modern genetic molecules RNA and DNA evolved. A popular idea is that a relatively simple 'RNA world' evolved first, but we are a very long way from understanding how that happened.

Lateral gene transfer still happens to some extent. Three mechanisms are known. Free DNA molecules can be absorbed by a cell. Cells can 'conjugate'—come into contact and transfer some DNA, a microbial precursor of sexual reproduction. A microbial virus can transfer DNA between cells. These processes now are severely limited by the protective mechanisms possessed by cells, and by the need for a transferred gene to fit into the biology of its host. When it happens it can be a problem for us, as with the transfer of antibiotic resistance between bacterial strains or species.

The initially anarchic community, Woese envisages, 'somehow pulled apart into two, then three communities, isolated by the fact that they could no longer communicate with one another in an unrestricted way'. The three superkingdoms were born. He thinks that at that time true organisms still had not evolved. Lateral transfer of genes and biochemical components was still frequent, though at a lesser rate than in the last common ancestor. DNA was already the main informational molecule of life, as it is used in all three superkingdoms. Soon within each superkingdom community the cellular systems became so integrated and distinctive that communication between them became highly restricted. Well-defined organisms emerged. This is highly speculative, but we can expect more insights as current research progresses.

STARTING CONDITIONS

Accretion of Earth continued until 3800 million years ago. Large impacts continued until that time (if we accept the presently poorly constrained interpretations). Many impacts would have been larger than the one that killed off the last of the dinosaurs, and the heat released would have vaporised the upper parts of the ocean.

If life started during this bombardment it is likely to have been extinguished. This could have happened many times. One proposition is that life arose at low temperatures, colonised the oceans including hot springs in the depths, and then all surface life was wiped out by an impact. This left the hyperthermophiles from the springs to recolonise the Earth, giving us a tree of life with hyperthermophiles at the base.

Much of the water in the oceans was delivered by comets from the outer parts of the Solar System, probably during the first 100 million years of the formation of Earth. Some components of the atmosphere would have had the same source, and along with more water some would have been 'degassed' from Earth, that is, released from rocks as they melted deep in the Earth and vented through volcanoes and springs to the atmosphere.

Knowledge of the climate of Earth when life arose would provide important constraints on conjectures about the origin of life. Was it hot or cold? Some 'prebiotic chemists' favour a cold environment for the origin of life, others a hot one. Cold environments are best for the preservation of a rich diversity of volatile organic compounds. Current understanding of the evolution of stars tells us that about 4000 million years ago the Sun would have given off 25–30 per cent less heat than now. Other things being equal, the whole Earth would have been frozen. But we know from the presence of the water-laid sedimentary rocks more than 3800 million years old in Greenland (chapter 3) that this was not so. The interior of the Earth at that time would have been hotter than now due to the accumulated kinetic energy from the infalling meteorites, and from energy released by radioactive decay of elements which now are less abundant. As a result there would have been abundant volcanism which pumped large amounts of carbon dioxide into the atmosphere, causing a powerful greenhouse effect and keeping the planet warm.

Early models of the Earth's primitive atmosphere were that it consisted of methane, ammonia and hydrogen. In famous experiments in the early 1950s American chemists Stanley Miller and Harold Urey simulated lightning discharges in such a gas mixture and were able to synthesise organic compounds, including amino acids. They were preceded in this work by the German chemist Walter Löb, who also synthesised amino acids, in 1909. More recent work, though, suggests that the atmosphere was mostly carbon dioxide and nitrogen. Methane and ammonia are destroyed by ultraviolet radiation in the upper atmosphere. Nonetheless, some methane was probably present in volcanic gases.

Given what we can infer about the accretion of the Earth, it is reasonable to conclude that life as we know it began no earlier than about 4000 million years ago. The Earth was its present size, but the rate of meteorite and comet bombardment was much higher than now. There were only small continental masses, but large

59

numbers of volcanoes. There was a global ocean, already very deep. Many of the volcanoes and minicontinents projected above the ocean, which was slightly acid, and hot. The solid Earth was hotter than now, from the leftover energy of accretion and the decay of radioactive elements. Hot springs were abundant on land and on the ocean floor. The springs and volcanoes emitted carbon dioxide, carbon monoxide, water vapour, methane, hydrogen, hydrogen sulfide, and other gases. The atmosphere consisted mostly of carbon dioxide, nitrogen and water vapour. It was probably very cloudy.

ABUNDANT ORGANIC CHEMICALS

Where in this tropical, acid, volcanic, bombarded environment were the carbon compounds that are the building blocks of life? Hypotheses of the origin of life fall into two groups—those that start with pre-existing organic compounds, and those in which the synthesis of the compounds is part of a continuum leading to life. There are many ways in which such compounds form, and not all, or even most, of the compounds involved in the origin of life had to have been made on Earth.

Interstellar space contains very dilute gas, mostly hydrogen. In places there are dense clouds of gas, and these are veritable factories for the production of organic molecules. In these clouds, a molecule will collide with another molecule, atom or ion about once a month (this gives us an insight into what astronomers mean by the word 'dense'!). On a time scale of 100 to a million years, these collisions build complex molecules. We can detect the molecules by the distinctive frequencies of energy emitted during collisions, which can be observed in telescopes. They can also be detected because of their absorption of starlight.

More than 100 different molecules have been identified in interstellar clouds and around 'red giant carbon stars'. Most are organic molecules. Those confidently identified have up to nine carbon atoms, but there is evidence for the presence of polycyclic aromatic hydrocarbons with up to 60 carbon atoms. As well as gases, the clouds contain solid particles of silicate minerals and carbonaceous matter, a hundredth to a tenth of a micrometre wide. Molecules condense on the particles as ices.

Some clouds collapse to form stars, with the distant parts of the clouds condensing as comets. High temperatures in the inner parts of new planetary systems, where rocky planets accrete, prevent the condensation of water, organic compounds and other volatiles. As the planets cool, comets serve as delivery vehicles bringing in these volatile compounds from the cold, outer parts of the planetary systems. It is thought that Comet Halley has organic compounds equivalent in weight to one-tenth of the biomass of Earth. In our own solar system, meteorites called carbonaceous chondrites also contain organic compounds, including amino acids, which they deliver to Earth's surface. It is reasonable to postulate that at a late stage in the accretion of the Earth, 3800 to 4000 million years ago, great quantities of such compounds were delivered to and incorporated into the uppermost crust.

It is thought that critically important starting products for 'prebiotic synthesis', the chemical processes leading towards the first living cells, are formaldehyde and hydrogen cyanide. Formaldehyde is needed for making sugars, and hydrogen cyanide for making amino acids and the bases in the centrally important molecules RNA and DNA. Both formaldehyde and hydrogen cyanide occur in interstellar dust clouds. The origin of the 'ribose' sugar in RNA and DNA, though, is one of the outstanding problems for prebiotic chemists, as attempts to synthesise sugars produce numerous different kinds in mixtures, and why life uses only ribose is not obvious.

So the starting point for some hypotheses of the origin of life is organic compounds ultimately derived from interstellar dust clouds. Other hypotheses emphasise the possibility of synthesis of these compounds here on Earth. For instance, both formaldehyde and hydrogen cyanide can be manufactured in a carbon dioxide–nitrogen atmosphere in the presence of trace amounts of methane. Light acting on carbon dioxide and water vapour can make the former, and lightning the latter.

The availability of organic compounds is unlikely to have been a major constraint on the origin of life, or on hypotheses of its origin. Our starting point then is an Earth at least locally rich in organic compounds, including amino acids, the building blocks of proteins.

AN ARRAY OF HYPOTHESES

Having garnered the basic chemical building blocks of life from whatever source, subsequent prebiotic chemistry can proceed in many different ways. Cells live as autotrophs or heterotrophs, or both. That is, they either manufacture their organic compounds themselves, starting from a simple compound such as carbon dioxide (autotrophs), or they assimilate them from their environment and make them into what they need (heterotrophs). Whether heterotrophs or autotrophs came first is one of the great issues that is debated. The heterotrophic origin hypothesis elaborated by the Russian chemist Alexander Oparin during the 1920s has dominated thinking this century. It is reasonable to postulate that the chemistry leading to life exploited the organic compounds already present, rather than starting from scratch. But more and more scientists now think that autotrophs came first.

One version of the autotrophic origin hypothesis has

been formulated quite recently. It is based on the discovery of the 'deep, hot biosphere', microbial ecosystems occurring deep underground. Some of the microbes are heterotrophs, but there are also autotrophs utilising hydrogen gas to reduce carbon dioxide to make cell material. There is some controversy about the source of the hydrogen, but in any event all the requirements of life can be found in this environment, and indeed life could have started there. This version is appealing because the environment is a sheltered one where the earliest microbes could have survived the last stages of bombardment of the Earth.

The German biologist Otto Kandler, one of the pioneers in the study of Archaea, points out that a great many hyperthermophiles are able to reduce carbon dioxide to make their cell material. That is, they are autotrophs. The energy to drive this process is derived from chemical reactions such as the reduction of sulfur by hydrogen. None is known to use the energy of sunlight—none is a photosynthesiser. He deduces that this indicates an autotrophic origin of life, but that 'photoautotrophy' is a later invention. In his words 'life was incited in the heated hydrosphere of a primaeval inorganic world by CO_2-reduction and organic synthesis driven by inorganic redox energy and electron flow'.

The molecular phylogenetic studies discussed in chapter 2, that led to the mapping of the tree of life, have revealed that all the known organisms close to the last common ancestor are capable of autotrophy. Heterotrophy seems to be an ability scattered randomly amongst these microbes. A reasonable inference would be that autotrophy is the ancestral characteristic, and that heterotrophy evolved later and spread by lateral gene transfer.

There has been a large amount of research on processes by which the simple chemical precursors discussed above could assemble into the much larger molecules found in cells, and into how the membranes characteristic of cells might first have formed. I can only provide the briefest glimpse of this work, with a few intriguing observations. Consider the linking together, polymerisation, of amino acids to build proteins. This has proven one of the most difficult processes to emulate. However, it has been found that, at about 60°C, dry mixtures of amino acids and phosphates will self-assemble into chains several hundred amino acid units long. Perhaps this once happened in evaporating pools on the flanks of volcanoes on the early Earth. In those same pools there would have been abundant clay minerals derived from the weathering of the volcanic rocks. The surfaces of clays are known to act as templates which catalyse the production of large 'polymeric' molecules from simple precursors. The clay kaolinite can catalyse the formation of long chains of amino acids, when the amino acids are first joined to a compound called adenosine monophosphate (AMP). This same linking of amino acids to AMP is employed by all life in the synthesis of proteins.

When some organic molecules that do not dissolve are dispersed in water they form hollow droplets called coacervates. These can be just micrometres in diameter, comparable in size to microbial cells. As with biological membranes, they can be multi-layered. They can form bud-like projections, and divide by pinching in two. They can contain other organic compounds concentrated from the surrounding medium. It is not difficult to imagine that just such a process could have been involved in the origin of the first cell-like structures. The coacervates could have incorporated the 'proteinoids' resulting from catalysis in those evaporative, clay-rich ponds described above, and the rich assemblage of other organic molecules found on the early Earth. The coacervates became reaction vessels, containing a rich broth of chemicals. Perhaps then higher levels of organisation evolved spontaneously in some of them, because of the myriad of reactions possible in such a concentrated brew. They were 'self-organising' systems. Whether we would call them alive is really only a matter of definition. Biologists consider that an essential characteristic of life is heredity, the ability to reproduce innovations. Nowadays this is mediated by the highly complex molecule DNA, but as discussed below, there must have been a simpler system to start with. The point at which a system can accommodate change and reproduce itself at least approximately, and therefore evolve, is the point at which we can talk of life.

Much more research has been done than I can even begin to describe here, and there is real hope that many processes potentially involved in the earliest stages of the evolution of life will eventually be understood. An appealing way to make progress is to read the tree of life literally, and then to apply Occam's Razor (that is, to prefer the simplest available explanation). The lowest branches of the tree are occupied by hyperthermophiles. The simplest interpretation is that life arose at high temperatures and then spread to cooler environments.

HYDROTHERMAL HYPOTHESIS

Natural springs of hot water are like an apothecary's kitchen of old, or the witches' cauldron in Macbeth '. . . Eye of newt, and toe of frog/Wool of bat, and tongue of dog . . .' They are rich in thermal and chemical energy, and are laced with nutrients leached from the rocks through which the water was passed. A number of scientists have proposed that life arose in such an environment. The idea has a history going back at least to 1924 and the American biologists R.B. Harvey and later J.J. Copeland, who studied the biota of the hot springs of Yellowstone National Park. It got a great fillip with the discovery in 1977 of the deep submarine 'black smokers' and their distinctive biological communities. These communities do not depend on photosynthesis as their ultimate source of energy and organic matter. Instead at their base are bacteria that gain their energy for growth from the oxidation of hydrogen sulfide emerging from the springs. These are the primary producers providing food to weird communities of giant clams and tube worms.

The existence of these communities and the fact that submarine springs are abundant sources of energy in the form of heat and chemical energy and essential nutrients in the form of dissolved chemicals rekindled interest in the notion of hot springs being the cradle of life. John Baross of the University of Washington and Sarah Hoffmann of Oregon State University were early advocates of this hypothesis. The water that emerges in hot springs began its journey as seawater seeping down through the sea-floor. The deeper parts of the oceanic crust of the Earth are very hot, and the water is heated to high temperatures, but does not boil because of the high pressures. As the hot water seeps through the basalt that makes up much of the deep ocean floor it leaches out metals and other elements. When it encounters fractures in the sea-floor it re-emerges as springs with temperatures as high as 400°C. Currently the mass of the oceans is circulated through the oceanic crust every 10 million years, and on a hotter early Earth it would have been even faster. Hydrothermal flow at temperatures less than 200°C is ubiquitous in the deeper regions of the oceans, while high intensity activity, with high flow rates and temperatures of 200–400°C, is abundant on mid-ocean ridges and in areas of volcanism. About 139 fields of these high intensity springs are known so far, and there could be as many as a million of the lower temperature fields. Such systems would have been even more abundant on the early Earth, when the heat flow was higher.

Chemist Everett Shock and colleagues from Washington University in Missouri and geologists Michael Russell and Allan Hall from the University of Glasgow have developed these ideas into a sophisticated hypothesis explaining what may have been the earliest steps in the emergence of life, steps they suggest could have happened in just weeks or months. At the end of this short period the first 'protocells' would have existed. These would not yet have been Carl Woese's progenotes, because they would have lacked RNA and DNA, but they were getting close.

'Energetically, hyperthermophilic anaerobes in present-day hydrothermal systems lead an extremely easy life', according to Everett Shock. 'Geological and geochemical processes supply them with abundant energy in easily used chemical forms . . . Tapping even a small amount of this energy leads to a rate of biological productivity which may be the highest on the planet.' What is more, using this energy to synthesise their cells releases even more energy. As Shock says, these may be the only organisms which can eat their lunch and have it too. This same energy can drive the synthesis of organic compounds in hydrothermal systems. Theoretical calculations show that as the hot fluids equilibrate with seawater, bicarbonate in the seawater is partly converted to organic compounds such as carboxylic acids, alcohols and ketones. At around 100°C, where hyperthermophiles thrive, synthesis of carboxylic acids is favoured. These are the basic constituents of cell membranes and other biomolecules.

Building on these calculations and their own observations of ancient hydrothermal mineral deposits, Michael Russell and Allan Hall have developed an elaborate

67

Figure 4.1 One of the many hypotheses for the origin of life. This version, taken directly from a research paper by Russell and Hall, has life emerging in a thermal spring on the ocean floor.

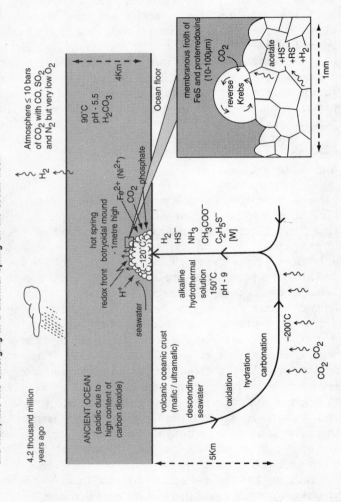

Source: *Journal of the Geological Society of London, 1997.*

hypothesis for the origin of life. Its key elements are shown in Figure 4.1 taken from one of their recent articles. The story goes like this: four billion years ago the ocean was slightly acid, because of the high content of dissolved carbon dioxide derived from an atmosphere rich in that gas. The ocean was also hot, because of the strong greenhouse effect caused by the carbon dioxide. The crust of the Earth was less oxidised than now, causing waters circulating through it on the ocean floor to be quite strongly reduced and alkaline. These fluids emerged on the ocean floor in many springs, some in the temperature range of 100–200°C where organic compounds are synthesised readily. Seawater and the intruding hydrothermal fluids were 'far from equilibrium' with each other. As demonstrated experimentally, where such waters mix, metal sulfides will precipitate in tiny bubble-like forms. The skin of the bubbles is predominantly iron sulfide, and they are membranous cell-like structures (the site of an emerging sulfur biochemistry according to Günter Wächtershäuser). The fluid inside the bubbles is a mixture of hydrothermal fluid and seawater, whereas that outside is entirely seawater. Thus the sulfide membranes separate two fluids still way out of chemical equilibrium. Equilibration is approached by flows of protons and electrons through the sulfide membrane. Bicarbonate from seawater also flows through the membrane and inside the bubbles is synthesised into organic compounds, in the way that Shock has predicted. This is the first, primitive, form of a biochemical cycle. It turns out that these deep ocean vents are also effective sites for the manufacture of ammonia, a building block of amino acids and nucleic acids. In experiments simulating the environment in the sea floor beneath the vents, Jay Brandes and colleagues in the Carnegie Institution in Washington found that at 500°C

governed by the Second Law Of Thermodynamics, is often regarded as inexplicable. The second law has the Universe degrading to the lowest common denominator, the minimum state of energy, the least order, randomness. Against this imperative we want to make RNA and DNA, veritable encyclopedias of information, miracles of complexity, the core of life. Information is not created from nothing: it flows between systems. But maybe we are even more ignorant than we realise. Perhaps there are fundamental laws of Nature still to be discovered. The theoretical physicists' quest for the unifying theory of everything is not the only example. There are other Holy Grails, and for life, it is the origin of complexity. Perhaps it will be found by the likes of Stuart Kauffman, or maybe the answer lies buried in the quantum properties of matter, as tentatively suggested by Paul Davies, and others. No-one knows. Davies thinks that the answer may lie in processes which operated at the very origin of the Universe:

> from the initial uniform cloud of gas, stars and galaxies congealed. Order appears spontaneously. In informational terms this seems all back to front . . . In some as yet ill-understood way, a huge amount of information evidently lies secreted in the smooth gravitational field of a featureless, uniform gas. As the system evolves, the gas comes out of equilibrium, and information flows from the gravitational field to the matter. Part of this information ends up in the genomes of organisms, as biological information.

> Paul Davies, *The Fifth Miracle*, Allen Lane, The Penguin Press, 1998, p. 36

Only a physicist could have the breathtaking audacity to link the origin of the Universe to the information content of DNA! Clearly there is plenty of work for the next generation of scientists.

CHIRALITY

Many large organic molecules have complex, convoluted shapes and can exist in two geometric forms which are mirror images of each other (as if they were left-handed and right-handed but otherwise identical twins). This property is called chirality and the different forms are called enantiomers. When compounds such as amino acids are made in the laboratory, equal quantities of these 'left- and right-handed' enantiomers form. But in all life on Earth, all chiral amino acids are left-handed and all sugars are right-handed. The origin of this asymmetry is a mystery.

Some meteorites contain organic compounds, including amino acids. Many of the meteorites have been dated and are known to be more than 4500 million years old. The organic compounds are samples from the early Solar System, and pre-date the origin of life. The Murchison meteorite fell in Victoria, Australia, on 28 September 1969. It is a 'carbonaceous chondrite', a type of meteorite rich in organic compounds. It was quickly studied by organic chemists and found to contain amino acids, including examples not known to be produced by organisms on Earth. Very recently it has been demonstrated that amongst these latter compounds is at least one with a 9 per cent excess in the left enantiomer, evidence that there is a pre-biological way to generate 'enantiomeric excesses'.

There is one mechanism known to produce such excesses in otherwise equal mixtures of organic compounds. Light polarised in a particular way, circularly, can do the job. Such light has the same type of asymmetry as chiral molecules. The energy of light can destroy chemical compounds, as when curtains fade in the sun. When circularly polarised ultraviolet light interacts with chiral substances, it destroys one enantiomer faster than the other. It has been shown that such interactions can produce enantiomeric excesses of 10 per cent or more.

Recently, Australian astronomer Jeremy Bailey, using the Anglo-Australian telescope, has found that light from one region of the Orion nebulae is circularly polarised. This is a region of 'high mass star formation'. As such regions are known to contain abundant organic molecules, Bailey considers that they are 'promising locations for the origin of chiral asymmetry'. Thus the chirality characteristic of all life on Earth may possibly owe its origin to the physical processes of star formation, somehow later amplified by the evolutionary selection of particular enantiomers.

It is also pointed out by some scientists that the surfaces of clay particles can absorb organic compounds, and act as templates in arranging those compounds in an ordered way. This too can lead to the selection of one enantiomer over another.

AN INTRACTABLE PROBLEM

Every scientific theory of the origin of life involves a large number of steps. There is no getting around the fact that even if the process of the origin of life was fast, it was immensely intricate. Every intricacy has its own uncertainties. We can never know all these steps. But we can know that it could have happened. Then if we know that, we can predict more confidently whether it might also have happened on Mars, or anywhere else. This is a field in which we are still hugely ignorant, but also one in which a vast amount has been learned over the last 50 years. It is perfectly reasonable to conclude on the basis of our current knowledge that life could have arisen spontaneously.

But we are so ignorant of the processes involved that we cannot make a meaningful assessment of the probability of the origin of life. If only chance were involved, then life is so improbable that we are likely to be alone in the Universe, no matter how many Earth-like planets there may be. But if as argued by biochemist Christian de Duve 'life is a cosmic imperative', a natural emergent property in a Universe characterised by processes of self-organisation, then it will be abundant. Experience tells us that any belief in our own uniqueness is hubris, so we can fall back to the most conservative conclusion: there is nothing special about Earth, and there is likely to be life all over the Universe. Obviously that is just speculation, and it is the truth of that proposition that we wish to test by going to Mars.

Chapter 5

Meteorites from Mars: a case of wishful thinking?

On 6 August 1996, the Administrator of NASA, Daniel Goldin, held a press briefing in Washington to make a sensational announcement. A team of scientists headed by David McKay of Johnson Space Center near Houston had found evidence of former life on Mars. It had been preserved in Martian meteorite ALH84001 (the name means the first meteorite to be catalogued from a collection made in 1984 at Allan Hills in Antarctica). The press conference was packed and the reaction from the media was as would be expected. The story led the news bulletins and splashed across the front pages worldwide. It was a major news and current affairs topic for weeks.

Here at last was hard evidence that Mars once harboured life. This was the first scientific evidence of life anywhere other than on Earth.

Apparently the press conference was arranged in haste, because the story had been leaked to the newsletter *Space News* (there are lurid and Machiavellian explanations of the leak, which could be true, but I will leave the telling of that story to those who are better informed). Formal publication came on 16 August in the journal *Science*. At the news conference Goldin and some of the scientists involved were careful to state that the evidence was equivocal and controversial, but convincing when the different lines of evidence were considered together. In an attempt to achieve balance in the presentations, Bill Schopf of the University of California, an eminent expert on early life on Earth (as we have seen in chapter 3), was asked to evaluate the evidence. Schopf was very sceptical of many aspects of the interpretation, and reminded us of the late Carl Sagan's comment that 'extraordinary claims require extraordinary proof'. In the end, though, all the words of caution were ineffective and the message was conveyed by the hype. As Daniel Goldin said 'this is a day that may well go down in history' and he was 'thrilled and humbled by this project'. The media certainly got the message. CNN television, for example, cut to the business news partway through Schopf's cautionary comments; they had given us the news, who needed the caution? The level of excitement at the conference was extraordinary. President Clinton was moved to state, 'If this discovery is confirmed it will surely be one of the most stunning insights into our universe that science has ever uncovered'. Cynics noted that the NASA budget was being considered by Congress that month, and that NASA needed a few wins; but given all the circumstances, that might have been coincidental, and in any event was unlikely to have been a factor considered by the scientists who had worked on the meteorite for more than two and a half years.

This is a complicated story. It is worth telling chronologically because it provides insights into the way science sometimes works. But if you want to go straight to the conclusion, it is this: the evidence is unconvincing and few scientists consider that it sheds much light on the possibility of life on Mars. However, the lessons learnt during the study of the meteorite have sharpened the skills we need in the search for life.

THE SNC METEORITES

ALH84001 is just one of the more than 7500 meteorites that have been collected. Meteorites are rocky debris that constantly rain down on Earth from sources in the inner parts of the Solar System, mostly the asteroid belt. A staggering 40 000 tonnes of this debris reach the Earth each year, not counting the giant meteorites and comets that impact here every few tens of millions of years. Most of the thousands of 'shooting stars' we see at night are smaller than peas and burn up in the atmosphere. But some are large enough to survive entry through the atmosphere. In recent years many have been found on the Antarctic ice cap where dark-coloured rocks are readily visible. Not only are they conspicuous, but the movement of the ice sheet concentrates them in fairly small areas. Where the moving ice meets a barrier such as the Trans Antarctic Mountains it is deflected upwards, and then eroded away ('deflated') by the powerful Antarctic winds, leaving concentrations of meteorites scattered over the surface. The number of meteorites found in Antarctica is as great as in the rest of the world combined. This is where ALH84001 was found in 1984.

Twelve known meteorites are considered to have originated on Mars. A thirteenth was recently announced but has yet to be confirmed. They are known as SNC meteorites, after three of the discovery locations: Shergotty–Nakhla–Chassigny. There is a famous (apocryphal) story that the one that crashed to the ground in Nahkla, Egypt, killed a dog on impact, in 1911. With a total weight of 80 kg, the equivalent of just two bags of cement, these meteorites represent our only available samples of Mars. The largest of them weighs 40 kg.

Meteorite or comet impact on the surface of Mars is the only natural process capable of ejecting Martian rocks into space. An impact capable of doing so would

78

leave a crater at least 10 km wide, and there are many this size and larger. Only very coherent rock types such as igneous rocks could survive so large an impact and be launched into space. So all the Martian meteorites are basalts or their deep-seated igneous equivalents. As improbable as it might seem, theoretical studies of the trajectories of rocks blown off the surface of Mars indicate that about 3 per cent of them will collide with Earth within 20 million years.

The twelve meteorites that have been recognised are considered to be a single 'family' because all are very similar rock types; all but one formed much more recently than other meteorites; all have an unusually high water content and abundant oxygen-bearing minerals; and all have a distinctive mix of oxygen isotopes unlike that in other meteorites. The young ages indicate that they came from a planet with at least some residual tectonic activity, rather than from asteroids. The evidence that they originated on Mars is threefold: they must have come from a rocky parent body such as one of the inner planets; their oxygen isotopic composition is unlike that of rocks on Earth or the Moon; and trapped in mineral glass in one of them (EETA79001) are gases that chemically and isotopically match the distinctive atmospheric composition of Mars as measured by the *Viking* landers during 1976.

The SNC meteorites are samples from subsurface parts of Mars: they have not been exposed to intensive weathering or to much irradiation by cosmic rays. It is possible to calculate when they were ejected from Mars by measuring the abundance of isotopes of helium, neon and argon generated by exposure to cosmic rays during their travel through space, and adding this age to the time they have been on Earth. They represent at least five separate impacts by asteroids or large comets, from around 14.4 million to 820 000 years ago.

ALH84001 weighs 1.9 kg and is about the size of a potato (Plate 6). Its Martian origin was not recognised until 1993, and it is unusual in being much older (4.5 billion years) than the other eleven meteorites from Mars (170 million to 1.3 billion years). These are the times when these rocks crystallised on Mars. ALH84001 was ejected from Mars 14.4 million years ago and was in the Antarctic ice for 13 000 years. It consists largely of orthopyroxenite, which is an igneous rock comparable to basalt but which crystallised in an underground environment without erupting to the surface. The 4.5 billion year age is based on measurements of samarium and neodymium isotopes. It is also unusual among SNC meteorites in containing carbonate minerals in veins and pore spaces. They are magnesium, iron and calcium carbonates. They fill fractures which formed when the rocks were subjected to an earlier impact. The time of that impact is very difficult to determine because it is hard to date carbonate minerals. However, the carbonate has now been dated at 3.9 billion years old, based on ratios of rubidium/strontium and uranium/lead isotopes. It is critical to know this date, because it is that which tells us when there was life on Mars, if we believe the rest of the evidence.

McKay and his co-workers describe five lines of evidence suggesting that the meteorite contains evidence of life:

- The carbonate is in the form of 'globules' comparable to crystal aggregates known to be produced by bacteria on Earth.
- Nanometre-scale carbonate structures in the globules resemble fossil spheroidal, rod-shaped and filamentous bacteria (a nanometre is a billionth of a metre, just ten times larger than a single atom). (See Figure 5.1.)

80

Figure 5.1 *Top*: Carbonate structure in Martian meteorite ALH84001, as seen with a scanning electron microscope. The segmented thread-like structure has been interpreted as a fossil microbe—it is half a micrometre wide (i.e., half of one-thousandth of a millimetre) (NASA photograph).
Bottom: Barium carbonate crystal aggregate grown in a silica gel, in the absence of any organisms. It is one micrometre wide. (Photograph courtesy of Stephen Hyde)

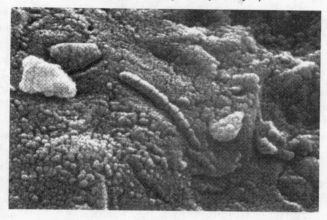

- Iron sulfide and iron oxide minerals are present in forms like those known to be produced by bacteria.
- The carbonate, iron sulfide and iron oxide minerals occur together but would not be stable under any one set of physical conditions, suggesting formation in the 'non-equilibrium' conditions characteristic of life.
- There is organic matter including complex hydrocarbons that could have been produced by living organisms.

All these features occur together in the carbonate veins. The researchers argue that these features are indigenous to the meteorite and 3.9 billion years old, and therefore that there was life on Mars at that time. They say that no one piece of evidence is compelling but taken together they are convincing.

The study of the meteorite was thorough but suffers from the limitations familiar to palaeobiologists who work with microbial fossils. As discussed in chapter 3, it is always difficult to find compelling evidence of a microbial origin for such simple structures which might equally plausibly be chemical and mineralogical artifacts (See Figure 5.1). Moreover, convincing fossils of this age are extremely rare on Earth, and very hard to find even in systematic searches of well-exposed regions of well-preserved rock. So it would be truly amazing if, with an available sample of just twelve rocks from Mars, one were to contain fossils. (Just recently, similar features have been reported from two more Martian meteorites, so my incredulity has reached new heights.)

LIMITATIONS OF THE INTERPRETATION

The initial report seemed to me, and to other scientists I have discussed this with, to be flawed in that it made

Top: The 'North Pole' area of Western Australia. These volcanic and sedimentary rocks are 3500 million years old and contain some of the oldest evidence of life on Earth. *Bottom:* Natural vertical section through a stromatolite at North Pole. During growth, this would have looked like a cabbage with the layers of rock recording former layers of microbes. The interpretation is controversial. The scale is 10 cm long.

Top: Stromatolites growing in 3–4 metres of water in Shark Bay, Western Australia. *Bottom left:* Shark Bay stromatolite sliced open to reveal sediment layers like those in the North Pole stromatolites. *Bottom right:* Microbes that participate in the construction of the Shark Bay stromatolites; they are thread-like cyanobacteria, the narrow examples of which are about two micrometres (two-thousandths of a millimetre) wide.

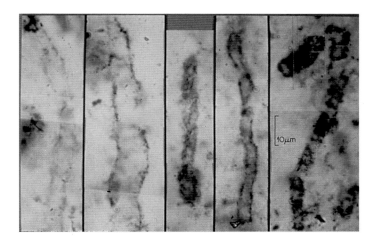

Top: Thread-like microfossils from North Pole, Western Australia, seen in 'thin sections' of chert viewed through an optical microscope. The microfossils are marked by dark remnants of the organic matter derived from the original cells, and are embedded within the chert. There is controversy about both the precise source and the age of these fossils. *Bottom:* Cyanobacteria from Mexico, which some scientists compare to the North Pole microfossils. The scale is 10 micrometres in both photographs. (Photographs courtesy of J. William Schopf)

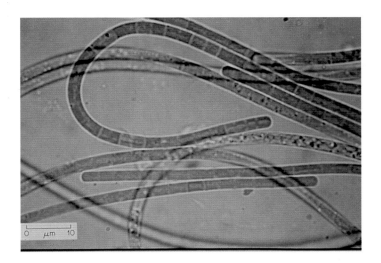

Thermal springs in Yellowstone National Park, Montana. *Top:* Grand Prismatic Spring. The pool is about 80 metres wide and is near-boiling. The zones around the pool are coloured by differently pigmented cyanobacteria, with those nearest the pool being the most heat-tolerant. The pool is in a mound of grey opal. *Bottom:* Castle Geyser, another mound of grey opal with coloured cyanobacteria.

Top: Artist's reconstruction of a typical scene on Earth 3500 million years ago. Volcanoes and thermal springs were prominent and stromatolites were abundant in the ocean. Parts of Mars might have looked very similar at the same time, or earlier. (Courtesy of the National Museum of Natural History, USA) *Bottom:* This scene in Shark Bay, Western Australia, mimics that of shallow seas 550 million years ago when animals first became prominent. A jellyfish swims above small seaweeds and mats of microbes. (Photograph by Steve Parker)

NASA/JSC

Top: Martian meteorite ALH84001. It is the size of an average potato. *Bottom:* The rover *Sojourner* analysing the Martian rock 'Yogi' during the 1997 mission, as seen from the lander *Pathfinder*. (Photographs by NASA)

Top: A scene captured by the camera aboard the *Pathfinder* lander on Mars during the 1997 mission. 'Twin Peaks' can been seen in the distance. (Photograph by NASA)
Bottom: A geological map of Mars. Different colours represent different types of rocks and terrains. This map by the US Geological Survey will be able to be upgraded after the current mapping mission by *Mars Global Surveyor*.

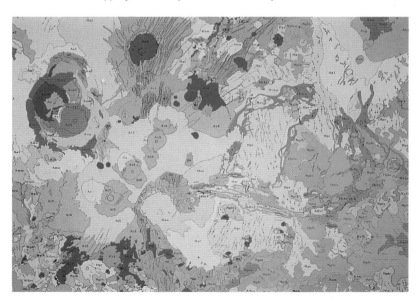

Mars as seen by the Hubble Space Telescope, 1998.

no effort to draw on the decades of experience palaeobiologists have in interpreting fossil bacteria. For instance, the article compared the carbonate microstructures to fossil bacteria, but made no attempt to make comparisons with non-biogenic mineralic structures that could be mistaken for fossil bacteria. There is a substantial literature on this topic. For instance, as mentioned earlier, Juan García-Ruiz has grown, entirely chemically, bizarre crystals—many of which look like microbes of various sorts. In the years following publication, several conferences have focused on the interpretation of ALH84001, and there have been many articles reporting new results. Neither the initial report nor any of the subsequent ones that I am aware of is by anyone with substantial experience as a palaeobiologist. Of course sometimes people without the inevitable prejudices resulting from familiarity with a field of research can bring fresh insights leading to breakthroughs. Often though they fall into traps long ago discovered by the old hands in the field.

The interpretation is that there was a community of disparate kinds of bacteria in the subsurface environment from which the meteorite was derived, and that at least several kinds can be interpreted by comparison with counterparts on Earth. So this argument reduces to a truly remarkable conclusion: not only was there life on Mars but there were several kinds of bacteria physiologically closely comparable to quite distinct kinds on Earth. The implication is that in the meteorite we have evidence of several twigs of the same tree of life that we know from Earth. This requires either that there were independent origins of life on Earth and Mars but that evolution was by some mechanism channelled in the same diverse directions; or that life arose on one of these planets and a community of bacteria was transferred to the other planet aboard a meteorite, as discussed later in this chapter (which must surely become less likely the more varieties of bacteria that are involved in each transfer event); or, and almost all scientists would regard this as implausible or even nonsense, both planets were seeded with life from some source outside the Solar System.

At the press conference the 'microfossils' were given most emphasis, although their interpretation was described as particularly controversial. In the published report they were compared to fossil bacteria such as those described in chapter 2, and to 'nannobacteria' reported from certain carbonate rocks on Earth. There are several serious problems with these comparisons: the style of preservation, the size, and the so-called Earth analogues. First, on Earth it is rare for bacteria to be preserved in carbonate minerals such as those in the meteorite, because such minerals readily 'recrystallise', growing into larger crystals and losing their original form. For this and other reasons there is very little known about preservation in such a medium, and so it is difficult to make convincing interpretations. Second, and more serious, the objects in the meteorite are one-fifth the diameter (one-fifteenth the volume) of the smallest known modern or fossil bacteria. The smallest 'nannobacteria' in the meteorite are 20 nanometres wide. Nearly all modern bacteria and all known fossil examples are 200 to 10 000 nanometres wide, in the micrometre size range, that is, some thousandths of a millimetre. To put this in perspective, if we enlarged a nanometre to the size of the head of a pin, then a metre would be more than 1000 km long.

Yet in support of their interpretation the authors are able to point to reports of the similarly sized 'nannobacteria' mentioned above. However, this is another major weakness in their argument, as the so-called 'nannobacteria' are not definitely bacteria. They are rock structures that one scientist in particular, Robert Folk from the University of Texas, considers to be fossil bacteria. Folk's interpretations are controversial because he has not been able to find any convincing evidence in support of his ideas. All he has is the shapes of mineral grains. There are no confirmed reports of nannobacteria. This is not to say that there may not be smaller bacteria yet to be found or reported, but the lack of known examples weakens the interpretation of the structures in the meteorite. There are reports of nannobacteria in human blood, but these are not free-living bacteria. They may be parasites of some sort, and their true nature is not yet understood. Even they are some 100 nanometres wide, five times larger than many of the features in the meteorite. Additionally, there must be a minimum size imposed by the metabolic requirements of life, and present indications are that it is somewhere near 50–100 nanometres wide.

Ken Nealson, an eminent microbiologist who recently joined NASA's Jet Propulsion Laboratory, has noted that some of the so-called fossil bacteria are so small that they could contain only one or maybe none of some of the molecules essential to life as we know it on Earth. Only the largest of the meteorite structures interpreted as bacteria are in the size range of known modern and fossil bacteria. Interpretation of the structures as the shrunken remnants of bacteria is not convincing either, as many examples of these are known as fossils and they are mostly readily recognisable and still fairly large. It could be suggested that they are fossil viruses, which are the right size, but viruses are not known to fossilise on Earth, and live only in or on other organisms. Any suggestion that they are fossils of 'ur-organisms', hypothetical precursors to bacteria or archaeons, would be regarded as an 'ad hoc' hypothesis, that is, a strained attempt to make the facts fit the desired interpretation.

But there is one very important positive point: the preservation of such tiny mineralic structures shows that if there were microbes on Mars then the conditions have been suitable for their long-term preservation as fossils.

As well as the 'nannobacteria' themselves, the carbonate globules in which they occur are intrepreted as possible bacterial precipitates. In terms of the shapes of the globules, that may well be a reasonable interpretation, as comparable concretions of calcium carbonate are known to be precipitated by bacterial action on Earth. However, such concretions also form non-biologically, and as McKay and his group admit, very little is known about carbonate petrography (crystal shapes and relationships) on this small scale. Consequently there is a limited database of comparable observations and thus, there can be little confidence ascribed to these interpretations.

The detailed chemical composition of the carbonates provides other clues. In favourable circumstances it is possible to deduce from the ratios of the isotopes of oxygen in carbonate minerals the temperature at which the minerals crystallised. When this approach was applied to the meteorite it was found that the carbonates formed at 0–80°C, consistent with the possibility that they formed in a near-surface hydrothermal system, just the sort of environment that could have harboured life. Subsequently, however, a study of the elemental composition of the carbonates suggested that they formed when carbon dioxide-rich fluids hotter than 650°C reacted with the rock during the meteorite impact which generated the fractures. Then along came a more detailed study of the isotopic composition of the carbonates employing an 'ion microprobe' that allowed in situ measurements of minute samples. This study found substantial variations of isotopic composition within the carbonate globules, said to be inconsistent with the isotopic equilibrium that would be expected in a high-temperature deposit. Using an entirely different approach, a group from the California Institute of Technology found that adjacent carbonate grains contain iron sulfide inclusions which record different magnetic field directions that were locked in when the minerals crystallised. They say that these different directions would have been lost if the minerals had been heated to high temperatures. Other studies add to the contradictions.

On Earth, evidence of life is also preserved in patterns of distribution of the isotopes of carbon, as discussed in chapter 3. In ALH84001, the carbonates are enriched in the isotope ^{13}C, consistent with the carbon source being carbon dioxide in the Martian atmosphere, which is similarly enriched now and perhaps also was when the carbonate formed. Some other SNC meteorites contain carbon with substantially different isotopic compositions, indicating that there has been isotopic fractionation of carbon on Mars. This can happen biologically, but there are also non-biological mechanisms which can produce the same results.

There is another isotope of carbon, ^{14}C, familiar to us as a result of its use in 'carbon dating'. It forms in Earth's atmosphere, by cosmic ray bombardment of an isotope of nitrogen, ^{14}N. It is short-lived, with a 'half-life' of 5730 years; that is, it spontaneously reverts to ^{14}N at a rate such that half has reverted after 5730 years. After 100 000 years the amount left is too small to measure. The carbonates in ALH84001 contain ^{14}C , which seems to indicate contamination since the meteorite arrived on Earth. There may be sufficient contamination to invalidate some of the organic analyses described later in this chapter.

Additional purported evidence for life comes from iron oxide and sulfide minerals in the meteorite. The carbonate globules have dark margins where there are abundant tiny crystals of the form of iron oxide called

87

magnetite. Comparable magnetite crystals are made by some bacteria which use them for orientation within Earth's magnetic field. These 'magnetotactic' bacteria have crystals with rounded and equant (blocky) shapes. Some of those in ALH84001 are very similar, leading McKay's team to use them as further evidence of life.

Later work has shown that there are also whisker-like crystals of magnetite which are thought to form only at temperatures of 500–800°C. The case against these crystals having been made by bacteria has been strengthened by the discovery of 'lattice defects' in some of them. These are defects in the pattern of stacking of the atoms making up the crystals. It is said that no such defects occur in magnetite made by bacteria, but this has been disputed. Most significantly, the so-called fossil bacteria could be explained as simply carbonate coatings on these magnetite crystals. One can also wonder why there would have been any magnetotactic bacteria on Mars when at present the planet has almost no magnetic field. However, it is likely that there was a significant magnetic field in the past, as indicated by remnant magnetism recently discovered (chapter 1).

There are also tiny crystals of the iron sulfide mineral greigite in the carbonate globules. These too are interpreted by McKay and his team as products of bacteria. On Earth there are bacteria that gain their energy for life by converting dissolved sulfate (sulfur with oxygen) to sulfide (sulfur by itself). The sulfide then reacts with any iron which is present to produce iron sulfide minerals. This is a very common process in oceanic sediments. However, there are also natural chemical processes that produce iron sulfides, so they need not be biogenic.

ALH84001 also contains organic compounds. They too have been misinterpreted, in the opinion of many scientists. Biological signals in terrestrial organic matter are preserved in two forms: specific organic compounds ('biomarkers') that can be traced back to components of the organisms that synthesised them, and distinctive carbon isotopic compositions that result from particular metabolic processes, such as photosynthesis. Organic matter in ALH84001, and also in the meteorite EETA79001, has carbon isotopic compositions indistinguishable from the commonest values found in terrestrial organic matter derived from organisms, which is depleted in the isotope ^{13}C. As Martian carbon dioxide seems to be highly enriched in ^{13}C these results could indicate a large amount of isotopic fractionation such as occurs on Earth in the biological production and consumption of methane. Or some of the organic matter could be an Earthly contaminant, as indicated by the ^{14}C analyses mentioned earlier. The results can also be explained non-biologically. Biomarkers should provide the answer. But none has been reported despite the application of very sensitive analytical techniques. What has been reported is the detection of a suite of aromatic carbon compounds, that is, compounds with the carbon atoms arranged in linked six-membered rings, called polycyclic aromatic hydrocarbons, PAHs. PAHs are common on Earth and result from the burning of fossil fuels and the natural thermal alteration of organic material in the Earth's crust. None occurs in organisms, and they are generally not regarded as biomarkers that can be related to organisms. There is organic matter, including PAHs, in many meteorites (especially the so-called carbonaceous chondrites), in dust particles from space (found in Arctic and Antarctic ice and collected at high altitudes by planes based on the former U–2 spy models), and in interstellar space (as detected and analysed by their infra-red absorption spectra). All this organic matter in space is taken as evidence for the sort of 'prebiotic' chemistry that might have led to the origin of life, as discussed in chapter 4.

In the paper in *Science*, McKay and his co-authors stated that the suite of PAHs in ALH84001 is unlike those known on Earth, and unlike those reported from interplanetary dust particles.

If the PAHs in ALH84001 are different to any other known PAH, then that difference can be suggested to be a biological signal. But are they really different? It seems not. It was suggested to me by Roger Summons, an eminent organic geochemist with the Australian Geological Survey Organisation, that the reported differences could be accounted for by the different analytical methods used. Subsequent studies of other meteorites, not from Mars, using the same methods, have revealed much the same suite of PAHs as in ALH84001. This means that these compounds are part of the rain of meteoritic debris constantly impacting the planets and tell us nothing about life on Mars. It is not even clear that the PAHs in ALH84001 came from Mars, because some later studies report the same compounds from Antarctic ice and it is possible that these compounds might adsorb onto carbonate minerals. This seems unlikely, though, because other Antarctic meteorites contain little or no PAHs.

A study by Hawaiian scientists Edward Scott, Akira Yamaguchi and Alexander Krot published in 1997 comes close to putting the final nail in the coffin of the biological interpretation. They believe they have demonstrated that the carbonates and other minerals in the veins crystallised at 200–500°C from minerals melted at the moment of the impact that blasted ALH84001 into space. Experimentally produced 'melts' of this type generate carbonate globules like those in the meteorite. If the fractures and the carbonate that fills them formed during a meteorite impact on Mars, or by some other high-temperature process, then the structures and compounds within them are extremely unlikely to be fossils.

A GRAND DREAM

What we are observing is the messy business of science in action at the frontiers of knowledge. In many ways there is nothing unusual about it, except that it is happening in a very public way. There is no doubt that the participants have been under enormous pressure both because of the potential significance of their observations and also because of the huge amount of publicity.

I was at a geological conference in Beijing when the news about the meteorite was announced. One eminent person with the requisite experience told me then that he did not believe the interpretations (at least as conveyed by CNN television to all our hotel rooms) but he would keep his mouth shut in the hope that something good might come out of the research in the end. The first reaction of several scientists I spoke to was to recall the 'cold fusion' controversy. This was the claim that nuclear fusion was achieved at room temperature in a laboratory at the University of Utah in 1989. This 'discovery' promised an unlimited supply of cheap energy. For obvious reasons it led to a huge amount of media interest, and substantial financial grants were made to capitalise on the initial results. However, the results have not been duplicated in any subsequent credible experiment. Some scientists still believe in it nonetheless. 'More than once, such believers have been intoxicated and misled by a grand illusion . . . Usually the illusion centers on some much-to-be-desired wish, like the cure of cancer' (*Cold Fusion: The scientific fiasco of the century*, by J.R. Huizenga).

This comparison with the report on ALH84001 has angered some scientists. There is absolutely no suggestion of any impropriety involved in the work on the meteorite. But the comparison is valid for other reasons: the initial report on 'cold fusion' was by scientists with no experience in nuclear physics, just as the meteorite researchers

91

have no experience in palaeobiology; the issue was hyped up enormously by the media; and it was considered to be in the interests of the home organisation of the scientists to maximise their credibility. The media are very good at making instant celebrities, and having once achieved stardom it is difficult to back down. Objectivity, always an elusive goal, becomes impossible.

Even when all the lines of evidence are combined as advocated by McKay and his team the result is unconvincing. So, unfortunately for those of us who consider that there may well have been life on Mars, it is easy to dispute the evidence from ALH84001. Evidence from meteorites is unlikely ever to be compelling, if we consider the difficulties of demonstrating ancient microbial life on Earth even when the full context of the samples is known. For meteorites there is no context. They are isolated pieces of rock. But as was hoped by my colleague in Beijing the intense burst of research on the meteorite is producing a lot of interesting new information. This will improve our ability to make more convincing interpretations of rocks collected from Mars, and on Earth. And it contributed to the continuation of research funding. An amazing array of innovative analytical techniques has been applied to ALH84001. Ironically, they have become part of the problem, because the new results cannot be interpreted by reference to a well-established body of knowledge. Most palaeobiologists have avoided the debate at least to the extent of spending time testing the observations and interpretations. That is because my scepticism is widely shared by my colleagues. We have learnt hard lessons in the past about misinterpreting evidence of microbial life.

One colleague said recently of the report published in *Science*, 'this is an article we had to have'. His thought was that the work raised both scientific and public awareness of the possibility of life on Mars to new heights, and stimulated a great deal of informative new work. Perhaps that is the most productive way to view the debate.

MICROBE TRAFFIC

There is another quite separate aspect of life in meteorites that has to be considered. That is the possibility—some scientists say the inevitability—that microbes could be transferred between planets aboard meteorites. Most bombardment of the planets by meteorites took place early in the history of the Solar System, until about 4 billion years ago. But large impacts have continued to occur sporadically, such as the one thought to have led to the extinction of the dinosaurs. Some of these impacts are energetic enough to eject rocks into space. It is estimated that every year, two tonnes of rock and dust from Earth rain down on Mars. Mars will be struck by a high-speed 1-km-wide object every few million years and this will be able to propel rocks outward from the planet at more than the escape velocity of 5 km per second. Most of the rock at the impact site will be vaporised, but some will survive, to be dispersed into the Solar System. It is estimated that each year the Earth intersects about half a tonne of Martian rock and dust. Comparable impacts occur on the other planets. There is a traffic of debris between the planets. The known cratering record and all plausible cosmogenic models show that this traffic would have been far greater during the first billion years of the Solar System. As we have seen, this was when life became established on Earth. Biologists have recently demonstrated that the biosphere on Earth extends down several kilometres into the crust, where there is a distinctive and extensive microbial ecosystem (chapter 2). It may extend to a depth of about 4 km on average, to where the temperature is about

125°C, near the currently known upper temperature limit for life. Estimates of the size of this ecosystem show that it must contain vast numbers of microbes, possibly equal in weight to all other organisms on Earth. Rocks blasted out of the uppermost crust of the Earth will inevitably contain samples of this ecosystem. So we are forced to consider the possibility of the transfer of microbes between Earth and Mars, in both directions.

The question becomes not whether there would be a traffic of microbes between the planets but whether they would survive the journey. The preservation of small-scale structures in the SNC meteorites proves that some rocks can be ejected into space with only mild shock damage. Theoretical modelling by Jay Melosh of the University of Arizona shows that around impact sites the shock waves can interfere in such a way as to produce regions where there are large accelerations but low shock. These are near the impacted surface. In a large impact the mass of lightly shocked rock ejected into space could be equal to as much as 1 per cent of the mass of the impactor. An impact which produces a crater about 50 km in diameter would eject a million cubic metres of boulders a metre or more wide from the lightly shocked zone where microbes might survive.

The travel time for meteorites between Earth and Mars will range from just a few years in extremely rare examples, to tens or hundreds of millions of years. There are many reports of bacteria surviving in a dormant state on Earth for millions of years, though this has yet to be convincingly demonstrated. Many kinds of bacteria are resistant to what to us are extraordinarily harsh conditions. Microbes inside a fair-sized rock, perhaps more than a metre or two wide, would be shielded from UV and most cosmic radiation. The temperature would be −10°C to −100°C which would not necessarily be fatal.

94

In fact, such low temperatures are beneficial for the preservation of bacteria.

Then the microbes would have to survive the impact on the receiving planet. Again, the preservation of fine structures and organic compounds in meteorites on Earth demonstrates that this is feasible. Having arrived at another planet, the microbes would have to propagate and disperse. Provided that the surface environment is habitable, as it was on both Earth and Mars early in their history, these seem to be the least of the problems.

Consequently, there is a potential explanation for any discovery of familiar types of microbes on Mars—they could have come from here. And the McKay team's contention that the features in ALH84001 are interpretable as remnants of bacteria like some known here could be explained by the interplanetary traffic of microbes. If life or fossils are ever found on Mars the first question that will have to be addressed is: are they descendants of travellers from Earth? This is not a new question: in 1871 William Thomson, later Lord Kelvin, in his Presidential address to the British Association for the Advancement of Science, said 'We must regard it as probable in the highest degree that there are countless seed-bearing meteoritic stones moving through space'.

You might be wondering what happened with the NASA budget in Congress after the announcement about the meteorite. It was cut by US$100 million. And the Administration proposed that another US$700 million be cut in 1998. But then in early 1997 President Clinton increased the proposed budget by US$500 million, to a total of US$13.5 billion, and the House of Representatives added another US$300 million on top of that. Perhaps this was because the US economy had improved, which it had, but it seems more likely to me that the President and Congress saw the huge amount of excitement generated by the possibility of life on Mars, and reacted to it. The quest for life elsewhere is inspirational, the leadership possibilities are extraordinary, and success would assure a place in history.

Finally, some trivia. There has always been a commercial market for meteorites, the more common of which sell for about US$100 per kg, but at a recent auction in New York tiny pieces of several Mars meteorites sold at prices of up to US$2000 per gram. I once met the science fiction writer Kim Stanley Robinson, well known for his novels about Mars. He told me that a friend had given him a unique and appropriate present—a tiny piece of a Martian meteorite. Guess what he did with it? He ate it, so he could call himself a Martian. So there is at least one.

CHAPTER 6
MARS ANALOGUES AND THE SEARCH STRATEGY

'From time immemorial travel and discovery have called with strange insistence to him who, wondering on the world, felt adventure in his veins . . . To observe Mars is to embark upon this enterprise; not in body but in mind.' So wrote American astronomer Percival Lowell in his book *Mars and its Canals*, published in 1906. And what a book it is. Building on the observations of nineteenth-century Italian astronomer Giovanni Schiaparelli, who mapped linear features he called canali (channels or canals), Lowell found many more of the same features and constructed a detailed and wonderfully imaginative interpretation. 'That Mars is inhabited by beings of some sort or other we may consider as certain . . .' On a drying planet these altruistic beings cooperated in an engineering feat of global proportions, building an intersecting network of canals to convey water from the poles to the arid equatorial regions. This is an extreme example of what the late American geologist Preston Cloud called 'geopoetry': in his conversations with me, Cloud described as geopoets those of his colleagues who constructed imaginative hypotheses for which, in his opinion, there was little evidence.

Lowell and his contemporaries were working at and beyond the limits of resolution of their telescopes. The features they drew on their maps were glimpsed at what they considered to be those rare moments when all the observational conditions were just right. It was meticulous and painstaking work. But we now have detailed images of Mars (Plate 8) and we know there are no canals, nor even, in most examples, features even roughly corresponding to those the astronomers drew. In Lowell's time the healthy scepticism of some fellow astronomers was not enough to overcome his enthusiasm and that of a public determined to believe in extraterrestrial life.

In the light of what has been learned over the last 30 years scientists long ago abandoned all ideas about advanced forms of life on Mars. But for many of the general public, the enthusiasm and determination still exist, and belief in the remains of a former civilisation on Mars persists.

There are many books about the exploration of Mars, and I have listed some at the end of this book, so I am not going to set out the history of the searches for life. After Lowell's time, the big events this century were the first images sent back by the missions of the 1960s and 1970s showing the planet to be an arid desert, experiments by the NASA *Viking* landers in 1976, conducted to search for living organisms on the surface of Mars, and imaging by the *Viking* orbiters. The focus was still on present life. Then with the *Viking* results of finding no life (at only two sites, we should remember), the focus in NASA shifted to searching for evidence of past life. The next great achievement came with the *Pathfinder* mission which landed the *Sojourner* rover in 1997 (Plates 6, 7). The possibility of there still being life on Mars is very real, and is still considered in the NASA program. It has even higher priority in the Russian program, and is a major component of that of the European Space Agency.

NASA

As set out in the next chapter, there have been 31 missions to Mars, launched by NASA; the space agency of the Soviet Union (subsequently Russia); and most recently, Japan. Thirteen were successful at least to some extent, and the result is a large and growing body of information about the planet. We now have detailed topographic and geologic maps, and a lot of other information, for instance on the climate. The spatial resolution of the data is generally of the order of hundreds of metres, which is good enough to allow a reconnaisance level of interpretation of the geology.

After a period of controversy about how to interpret the *Viking* results, most scientists concluded that they produced no evidence of life. In fact, they show us that the present surface is very inhospitable, with no liquid water, seemingly a chemistry so oxidising that no organic molecules could survive, the lack of an ozone shield in the atmosphere resulting in the surface being flooded with sterilising ultraviolet radiation, and the near lack of a magnetic field to keep out deadly cosmic radiation.

There is now detailed information about the *Viking* and *Pathfinder* landing sites, and a broad coverage of the whole planet. There is no sign of life. But that does not mean there is or was no life. The search has just begun.

Since the collapse of the Soviet Union the exploration of Mars has been dominated by NASA. In the competition for funding and for experiments on missions that characterises the internal workings of NASA, 'exobiology' fell behind after *Viking*. A small group of dedicated believers, including David DesMarais and Chris McKay, later joined by Jack Farmer, supported by some of the more visionary senior administrators, worked hard in an endless round of conferences and committee meetings to rebuild support for exobiology. In the middle of their struggle came the tragic loss of the space shuttle *Challenger* and all its crew. The resulting enquiry found unacceptable levels of cost-cutting and risk-taking in the construction of the shuttle, and the whole space program and NASA fell into disrepute. Over the following years the exobiologists continued their research on early life on Earth and their efforts in rebuilding their program. Then in 1993 came the loss of the US$1 billion dollar *Mars Observer* mission just as the spacecraft was about to go into orbit around Mars. The very existence of NASA was under threat.

Despite the serious problems, a core of strong public and professional support remained, organised for instance by the Planetary Society, and cosmologist Carl Sagan. As a result, exobiology has not only survived but has metamorphosed into 'astrobiology', and in 1998 the Astrobiology Institute was established within NASA. This has no normal physical presence: it is a number of research groups linked by the Internet, soon to be upgraded to a new high-speed version. The greatest threat to the program now is that the International Space Station, currently under construction, will divert all the funding to a program that many believe has more to do with politics than science.

LOGIC

The *Mariner* and *Viking* images showed us that the planet's surface was once hospitable, perhaps more than 2500 million years ago. So to find evidence of life the best chance is to go to rocks at least that old. How that search is conducted is determined by a small number of individuals who take their advocacy to the committee meetings from which the mission plans eventually emerge.

Search strategies are based on the one model we

have of what Martian life might be like—Earth. This approach is often criticised by non-scientists, but it is the way science works. Scientists make simplifying assumptions and then test them.

This does not mean that other possibilities are ignored, but we go for the main chance first. In the example of Mars there are three powerful reasons for taking this approach: all our current understanding indicates that life anywhere is likely to be carbon-based and require liquid water, and therefore be at least broadly comparable to Earthly life; there is a whole planet to explore, and limited resources, so we need sharply defined target areas; and, really a re-statement of the last point, there is not enough money to explore every possibility. This is not as single-minded as it might seem, because within the community of scientific explorers there are different points of view which compete for the resources of the missions, with the result that all the eggs do not end up in one basket. It does mean, though, that discredited ideas like the former presence of a civilisation on Mars get short shrift, and rightly so.

There were only microbes on Earth at the time that the surface of Mars was habitable. A number of major lessons have been learned from the search for ancient microbial life here, and these guide the search on Mars. One of overriding significance is the need for multiple criteria for biogenicity, as discussed in chapter 3. One line of evidence for proving that life was present is rarely enough. Ideally we want to find say, fossil cells, relevant carbon isotope patterns, biomarker hydrocarbons, and macroscopic structures such as stromatolites. The probability of finding such diverse evidence of life must be one of the selection criteria for search sites. Second, the chance of any particular sort of evidence being preserved in a particular type of site has to be considered. For instance, if we went to an ancient lake site on Earth what is the chance that we would find microfossils? How does that compare to the chance at a former hot spring site, or a fossil soil?

We can also look at the history of significant discoveries on Earth and reconstruct the logic, with the aim of applying it on Mars. Let's consider some examples. Perhaps the most influential discoveries of very ancient fossils on Earth were of fossil microbes 2000 million years old in the Gunflint Formation of Canada, 830 million years old in the Bitter Springs Formation of Australia, and 3500 million years old in the Warrawoona Group of Australia, stromatolites also in the Warrawoona Group, and very early animals 550 million years old at Ediacara in Australia.

The Gunflint fossils were found by American geologist Stanley Tyler when he was studying the rock types, not looking for fossils. It was fortuitous. He showed them to palaeontologist Elso Barghoorn of Harvard University. Barghoorn understood their significance and also learned an important lesson: black chert, a sort of flint, is good for preserving microbial cells. He set about travelling the world searching for chert. One way he did this was to visit government geological surveys and ask around. In Australia he met geologist Helmut Wopfner who told him about the Bitter Springs Formation then being mapped by geologists of the Bureau of Mineral Resources. So that discovery resulted from a systematic search grounded in regional mapping by government geologists. The Warrawoona Group stromatolites were a fortuitous discovery by John Dunlop, a student of the University of Western Australia studying other things in the area for his PhD; he recognised the significance of his discovery. He had a prepared mind. Others were found at about the same time by geologist Don Lowe of Louisiana State University during a study of volcanic rocks—a fortuitous discovery, and another prepared mind. The microfossils from the same area were found as a result of systematic searches by several pal- aeontologists, attracted to the area by knowledge gained from previous basic geological mapping and study by others. The Ediacara animal fossils were found by government geologist Reg Sprigg during a survey of a copper prospect, another example of a mind open to novel discoveries.

Why does Australia figure so large? Not just because it is where I am writing this book. There are large areas of ancient rocks here, they have been systematically mapped, and have been studied from many points of view. As a result there is a large basic database, and it is publicly available. Physical access is relatively easy; in these modern times there are no significant political or geographic barriers to inexpensive exploration.

The common threads here are availability of back- ground information, particularly geological maps; understanding of what to look for; systematic, focused searches; and openness to fortuitous discovery.

The search strategy involves detailed mapping of Mars (Plate 7), and the collecting of information on the composition of rocks, by satellite and unmanned rovers and probes. To refine the exploration process, sites on Earth considered to be analogous to targets on Mars are being studied (these include several sites in Australia). This has been going on for a long time, about twenty years. The initial focus was on Antarctica, particularly the 'Dry Valleys' where there is no ice. There was a lot of work on endoliths—microbes that live in the pores in rocks. The idea was that even on a dry Mars there might have been enough moisture for life within the rocks. Other work focused on the lakes in Antarctica as ana- logues for former lakes on Mars. Antarctica as a whole is an appealing analogue for Mars because both are frozen but one still teems with life. The work continues there, but now concentrates on using the lakes as test sites for remotely controlled vehicles. Interestingly, Antarctica is also being considered as an analogue for Europa, a moon of Jupiter. There may be liquid water beneath the icy surface of Europa, just as there is in 'Lake' Vostok, discovered under the Antarctic ice sheet during seismic surveys.

What sorts of sites on Mars might have fossils? The sediments of lakes and springs are being given high priority, and also the regolith (rocky soil) and the ice. The first two are the most promising and it is these I discuss below.

LAKES

Sediments forming the beds of ancient lakes are an obvious target. On Earth, lakes are a good place to live. Microbes of all sorts can be found in them, and many lakes contain stromatolites. When the microbes decompose in the sediment, they leave a record in the form of hydrocarbon biomarkers, and carbon and sulfur isotope patterns.

A number of possible former lake sites have been identified on Mars. Let's imagine that we have placed a lander in the middle of a dry lake bed inside an ancient impact crater 50 km wide. We have a rover with the ability to travel up to 5 km from the lander. On board the rover are cameras, a microscope, X-ray spectrometer/diffractomer for identifying minerals, a mass spectrometer for measuring isotopes, a gas chromatograph for detecting carbon compounds and a sample collecting device. All this is feasible. Let's say the lake resembles the great salt lakes of Australia, or Great Salt Lake of Utah. In the Australian examples, the chance of the cameras spotting stromatolites would be almost zero, because in most examples there are none. They would be higher if it was like Great Salt Lake, where stromatolites form around the margins, perhaps in zones of groundwater seepage. Finding them would depend on the rover being able to reach the edge of the lake bed. Then we would face the problem of determining whether the objects were stromatolites, or perhaps just chemical precipitates that formed around springs. On Earth, stromatolites occur in just about all well-preserved very ancient carbonate beds, but definitively recognisable examples are restricted to less than 10 per cent of randomly selected samples.

We would want to look for fossil cells, microfossils. So our sampling device would drill down 20 cm into the lake floor, to sample the hardened mud. We would like to drill deeper, but the solar cells that power the rover cannot generate enough energy to make that possible (and because of public opinion back home we were not permitted to fly a nuclear power generator). On Earth, if the rocks are fresh (not oxidised by weathering), as from drill holes or frigid, glaciated environments, up to 30 to 50 per cent of samples selected on the basis of fine internal particle size and rock colour yield fossils. However, most of these are fossil 'protists', usually eucaryotic algae which form very resistant and preservable cysts when times are tough. It took more than two billion years of evolution to get to that stage on Earth. Here we are on a lake bed three billion years old, and the chance of there having been such 'advanced' life is near zero. We are looking for the equivalents of bacteria and archaeons. Unfortunately these do not preserve well in mud. Nearly all of them decompose. A few have resistant structures that survive, but the shapes are not very distinctive and may not be obviously biological. If we were able to find some black cherts, experience suggests that 1 per cent of samples might contain microfossils. In a crewed mission, an experienced collector might be able to improve the odds to about 30 per cent, the best result on Earth.

Never fear, the gas chromatograph and mass spectrometer will find the evidence we need. Unfortunately, it turns out that because of the highly oxidising environment on the surface of Mars, no organic compounds are present in the sediment at shallow depths. There is some calcium carbonate, and we measure its isotopic composition. But we cannot make a convincing interpretation of the result because we have no organic carbon to contrast it with.

All pretty dismal, but quite likely. There could well have been life in the lake, but the evidence has eluded us. I advocate a different type of target.

SPRINGS

Here I have to admit my bias. I have been an advocate of targeting spring deposits on Mars since I was invited to a NASA conference in 1987 and asked to propose a search strategy.

There is one environment where more than 95 per cent of randomly selected samples yield morphological evidence of life—springs (Plate 4). Stromatolites are almost ubiquitous, and microfossils are very common (though often poorly preserved). It is fairly easy to prove that hot spring stromatolites are biogenic, because not only do they frequently have microfossils preserved within them, but the microfossils are systematically arranged in ways which allow their role in construction to be deduced. Springs not only are sites of prolific life, they are also sites of abundant deposition of minerals that entomb and fossilise the biota. Hot springs in particular support a wide range of metabolic strategies— every major biochemical strategy for life can find a niche in this environment. Bacteria and archaeons are abundant

Figure 6.1 **The 45 km wide outflow channel Dao Vallis (medium arrow) is thought to have been eroded by water flow from springs on the flank of the volcano Hadriaca Patera (large arrow). The small arrow indicates small channels. (NASA *Viking* image from Farmer (1996) in *Evolution of Hydrothermal Ecosystems on Earth and Mars*, John Wiley & Sons.)**

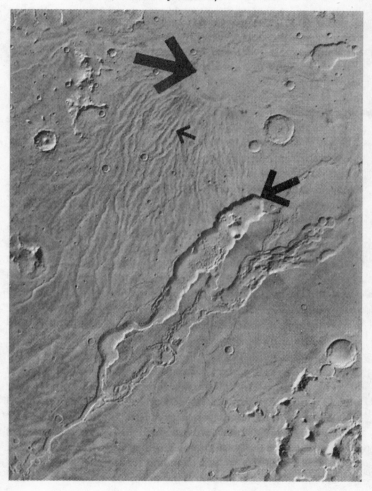

and diverse. And as discussed in chapter 2, this environment may have been the cradle of life on Earth. It is just the sort of place to look for life as it may have been 3 to 4 billion years ago.

A disadvantage of many springs is that they seem to be poor repositories for chemical evidence of life. In many examples, the deposits that form are very porous, and most organic matter is rapidly oxidised away. There are some exceptions to this, particularly where the sediments were deposited under stagnant water. Very little research has been done on such spring sediments. One ancient possible example that is famous for its preserved microbes, plants and insects is the Rhynie Chert of Scotland. This 400 million year old deposit has been known since early this century. It has long been considered to be a thermal spring deposit, but it has been difficult to prove that interpretation. The problem is that the deposit does not outcrop, and most of the work that has been done on its fossils has used samples collected from the soil in a farmer's field, or from the stone wall of the field. Recent work by geologist Nigel Trewin and his colleagues at the University of Aberdeen has confirmed the thermal spring interpretation. Trewin had access to drill core obtained when the deposit was explored for gold. The black cherts that contain exquisitely preserved fossils may have formed in short-lived swamps surrounding the springs.

The Rhynie Chert not only contains fossil cells but also preserved organic matter derived from the original organisms. In this example the full range of palaeontological tools can be applied, including isotope and biomarker geochemistry. In contrast, slightly younger spring deposits in central Queensland that I have studied with colleagues from NASA and the University of Montana contain very little organic matter. These deposits at Wobegong and Verbena are pale-coloured cherts.

Stromatolites and microfossils are abundant. It is not clear whether the lack of organic matter is the result of relatively recent weathering or very early decomposition. Thermal spring sediments I have studied in Yellowstone National Park in Montana, and in the Rotorua area of New Zealand lose most of their organic matter quickly when springs change and deposits dry out. This is because they are very porous and air gets into them readily, providing oxygen to destroy the organic matter. The same would happen on Mars, because although there is not much free oxygen, the surface is thought to have highly oxidising peroxides.

An advantage of targeting spring deposits is that suitable exploration techniques are highly developed because hydrothermal systems (of which springs are the upper part) are sites of formation of many valuable ore deposits—gold, silver, copper, lead and zinc in particular. The Queensland springs I have studied were discovered during the exploration for gold. They are underlain by low-grade gold deposits, and others nearby are being mined. Exploration geologists make a living out of finding them, so the technology is highly developed. Two of the more successful techniques are looking for the characteristic mineral signals on satellite images, and predicting where such deposits might lie using models, hypotheses, of how they form. Both these techniques would work equally well on Mars, yet the expertise of exploration geologists is barely being utilised at present.

The hot spring strategy is now very popular. There is evidence of springs on Mars in images from the *Viking* orbiters, tentative recognition of hydrothermal clays (nontronite), and in the mineralogy of Martian meteorites. They are inevitable, given the presence of water and heat (volcanoes). Climate modelling has recently suggested that liquid water would not have been stable even 3–4 billion years ago on Mars, and if that is correct, then

109

hot spring deposits could be the only promising target for finding fossils.

ORE DEPOSITS

So now instead of looking for the proverbial needle in a haystack, or in our case a microbe on a planet, we can focus on looking for large targets—lake and spring deposits. With springs, we can think of the job as exploring for ore deposits. What at first might have seemed a near-impossible task is thus reduced to an activity that is an everyday job for exploration geologists. Being an optimist, I think there might be a way to make it even easier.

What if there were a type of ore deposit where microbes played a critical role in its formation? So critical that without microbes there would be no ore? Then merely locating such a deposit would be evidence that there had been life at work, without lots of fancy analyses having to be done. We could just fly our satellite over Mars with spectrometers tuned to detect the right combination of minerals, find a deposit, and bingo!, evidence of life. Well, it is never going to be so easy, but the idea has merit. On Earth, microbes have profoundly altered the chemistry of surface environments, and probably also those of the upper few kilometres of the crust. For instance, the change from oceans and an atmosphere with almost no free oxygen to a situation approaching the oxygenated world we have now was brought about by the evolution of oxygen-releasing photosynthesis by the cyanobacteria, at least 2700 million years ago. Any mineral deposit that forms in habitable environments has the potential to carry a signal of the presence of life. In fact it is inevitable. The trick is recognising the signal and proving it to be biological.

Consider the HYC silver-lead-zinc deposit at McArthur River in northern Australia. The metals are

110

present as very thin layers of their sulfides in rocks that originally formed as muds and silts in a submarine environment. About 1640 million years ago hot metal-bearing waters moved up a vertical fault and near the sea-floor spread out into the recently deposited sediments. By some mechanism, sulfate dissolved in the waters was reduced to sulfide and precipitated the metals, over an area in excess of 3 km by 2.5 km. The ore deposit is rich in evidence for life. Black cherts that formed during ore deposition contain abundant microfossils. Carbon isotopic evidence of photosynthesis is ubiquitous, as is sulfur isotope evidence of bacterial sulfate reduction. Every sample of the ore, even if collected by a dumb robot, would display this evidence. Also ubiquitous are biomarker hydrocarbons. These demonstrate the former presence of bacteria, including forms that oxidise sulfur, archaeons, and algae. This ore deposit is a palaeontologist's dream, as well as being one of the largest silver-lead-zinc accumulations on Earth. If we could find such a deposit on Mars, the game would be over.

The conventional interpretation of the many deposits like the HYC is that the sulfide that precipitated the metals was produced by purely chemical reduction of sulfate when it reacted with organic matter in the sea-floor sediments. The evidence is that this happened at about 100° to 150°C. However, no-one has ever been able to reproduce this reaction in the laboratory at temperatures less than 200°C, despite numerous attempts. Nonetheless, most scientists have assumed that it must be possible, because the only other way to do it would be by bacterial sulfate reduction. This is a very common process, but 'everybody knows' that it could not possibly occur at more than 100°C. This is wrong, we now know. Neither the upper temperature limit for bacterial sulfate reduction, nor the upper limit for life as a whole, are yet known, but 150°C is widely believed to be a good estimate. Bacterial sulfate reduction did happen at HYC, as demonstrated by sulfur isotope analyses, but the anti-biological hypothesisers contend this was near the sea-floor and precipitated only iron, not deeper in the sediment where the silver, lead and zinc were precipitated. What happened to reduce the sulfate at HYC is not yet known, but I believe the evidence is consistent with the process being biological. If this is correct, and with several colleagues I am studying that possibility, then the whole ore body is evidence of life. Of course it is anyway, even if the chemical reduction hypothesis is correct, because then the process requires organic matter to be present in the sediment, and that is derived from organisms. No matter which hypothesis you accept, such ore deposits are evidence for life.

We could argue that on Mars even if there was never life, there could have been non-biological organic matter derived from meteorites and comets, and this could have reacted with sulfate to produce sulfide. We could also argue that there would never have been enough sulfate anyway if there had been no photosynthesis to release oxygen to oxidise the sulfide in the first place. What this points to is the need to generate for Martian conditions the sort of ore deposit 'models' that exploration geologists use on Earth. This is a standard exploration technique: the better we understand how ore deposits form, the better we can predict where they might be. Such an approach has not yet been attempted in the search for life on Mars.

Exploration costs a lot of money, and we are never going to be able to collect all the information we would hope for. So if we are to have any hope of answering the question as to whether there was once life on Mars we will have to finesse our way through a forest of uncertainties and inadequate information. Exploration geologists make a profession of doing just that, in their search for mineral deposits and petroleum accumulations. Using their experience and techniques, we can help design precise experiments and define precise targets on the surface of Mars (and indeed on Earth, in our search for earliest life here).

It is reasonable to ask whether the whole strategy might be wrong. Will picking winners work? Should the strategy at least include reconnaissance, for instance along transects? How about one mission with long-distance rovers to collect large numbers of very small samples? At present this might well be impractical for technical reasons and a good compromise is more missions like *Pathfinder* (chapter 7). Here the strategy is to go to a place like an ancient river delta, or the debris from a large impact crater, to observe and sample rocks derived from a wide range of environments.

THE LANDING SITE CATALOGUE

Anyone with access to the World Wide Web can go to http://www.nasa.gov and find a travel catalogue for Mars, compiled by NASA. It is called the Mars Landing Site Catalog. In it are descriptions, maps and images of about 200 sites identified as of interest for a range of different scientific reasons. The maps even show recommended routes for rover exploration. About 50 sites were selected because they could contain evidence of former life. The other sites are of interest for the geology of the planet. As the current missions improve the mapping no doubt new sites will be added and others deleted.

We can get a feel for the sort of information that is available and the approach that is being taken by considering two sites, a lake and a spring system. What follows are direct quotes from the catalogue for two sites, Gusev Crater and Dao Vallis.

113

Site Name: Dao Vallis

Type of Site: Balloon/Small Rover

Latitude: 33°S
Longitude: 266°W
Elevation: +1.5 km

Maps: MC-28 NE; ASU-IPF 713

Viking Orbiter Images: 62S18-20, 32A529, 408S69 through 408S70, resolution is 108 m/pxl; 410S02, 410Sd04, 410S06, 410S09, resolution is 165 m/pxl; 124A34, 124A37 through 124A40, resolution is 145 m/pxl; 625A17 through 625A20, resolution is 235 m/pxl

Date Entered: 6 September 1989
Date Last Revised: 22 March 1991

Geology Contact:
  Frederick R. West
  519 Monroe Circle
  Glen Burnie, MD 21061-3928

Exobiology Contacts:
  Jack D. Farmer
  NASA-Ames Research Center
  Mail Stop 239-4
  Moffett Field, CA 94035-1000
  (415) 604-5748
  Fax (415) 604-1088
  E-mail: jack_farmer@qmgate.arc.nasa.gov

  Ragnhild Landheim
  Depts. of Botany and Geology
  Arizona State University
  Tempe, AZ 85287-1404
  (602) 965-7029
  Fax (602) 965-8102
  E-mail: landheim@asuws2.la.asu.edu

Geologic Setting

The upper western branch of Dao Vallis, an apparent outflow channel of older material (Hch), lies between Hadriaca Patera eruption (AHh) and smooth Hesperian plains (Hpl₃) materials. It drops 5 km in elevation from its spatulate depression origin on the southern slope of Hadriaca Patera to its terminus in Hellas Planitia's modified aeolian, fluvial, and lava flow material (Hh₃).

Scientific Rationale

Investigation of a channel, the western branch of Dao Vallis, the volcanic eruption and Hesperian plains material adjoining it, and possibly the lower Dao Vallis to its terminus in Hellas Planitia. This area contains at least four types of terrain units. There is the possibility to investigate frosts, water vapor, clouds, and dust storms both on the channel floor of the Dao Vallis channel and also in a part of Hellas Planitia.

Objectives

A Balloon Exploration Vehicle is to be landed in the upper end of the western branch of Dao Vallis.

After leaving a lander rover surface exploration vehicle at this landing site, the balloon will, when suitable northeast winds occur, take off for daytime flights southwest over the floor of Dao Vallis, landing on the channel floor at night for direct study of the channel floor. The balloon will make daytime close-range observations in flight of the channel and possibly of adjoining terrain. If the balloon vehicle successfully reaches the junction of the western and eastern branches of Dao Vallis at 37°S, 269°.8W, it should continue to fly southwest above the lower Dao Vallis channel and make observations during daylight hours, and follow Dao Vallis at least to its terminus in Hellas Planitia.

Potential Problems

Suitable wind to keep the balloon flying southwest over or near Dao Vallis. Cliffs at the edges of Dao Vallis channels; winds and air currents near them. Dust storms, haze, and clouds. Dust and rubble on channel floors.

Trafficability

*To be determined; see **Potential Problems** above.*

Estimated Traverse Distance

300 to 1,000 km for Balloon Vehicle. Up to 300 km for the lander rover.

Exobiology/Significance

The simple channel morphology and enlarged source area suggests that Dao Vallis may have been formed by spring sapping. Furthermore, the close proximity of Dao Vallis to a potential subsurface heat source (i.e., magma chambers underlying Hadriaca Patera) suggests the possibility of hydrothermal activity. The proposed landing site is located at 33.2°S, 266.4°W near the head of Dao Vallis, presumably in the area of the most recent activity. The exobiology target was chosen in the central part of the channel to avoid sampling wall debris.

Formation of Dao Vallis clearly post-dates the major constructive phase of the volcano, although the amount of time separating events is problematic. The simple form of the valley and its large size is interpreted to be more consistent with a prolonged period of hydrologic activity, than with a catastrophic release of water.

The proposed site is of great interest to Exobiology because of the possibility of sustained hydrothermal activity and associated mineralization. Using Earth-based analogies, thermal spring facies are regarded as excellent targets in the search for a fossil record on Mars because of the high biological productivity and pervasive early mineralization typically associated with such systems. However, high resolution visible range imaging is needed to understand the origin of smaller-scale features (such as the knobby terrain on the floor of Dao Valis) and infrared spectral data is needed to support the search for hydrothermal mineral deposits.

Site 032 DAO VALLIS

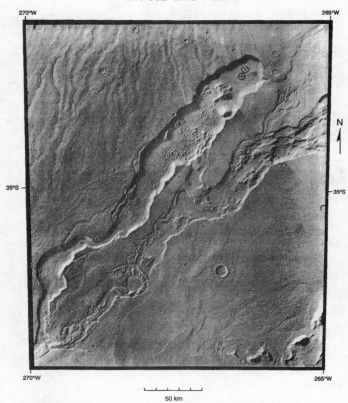

270°W   265°W

N

35°S

50 km

ASU -IPF 713   ⊗ Landing Site: 33° S Latitude, 266° W Longitude   F.R. West
                    ⊕ Exobiology Site: 33.2° S Latitude, 266.4° W Longitude   J.D. Farmer
                                                     R. Landheim

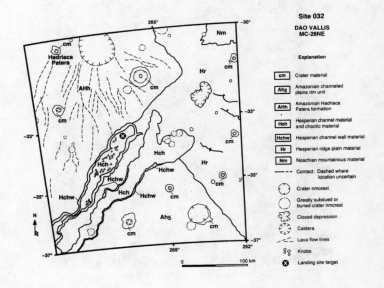

Site 032

DAO VALLIS
MC-28NE

Explanation

cm	Crater material
Ahs	Amazonian channeled plains rim unit
AHh	Amazonian Hadriaca Patera formation
Hch	Hesperian channel material and chaotic material
Hchw	Hesperian channel wall material
Hr	Hesperian ridge plain material
Nm	Noachian mountainous material

— — — Contact: Dashed where location uncertain

Crater rimcrest

Greatly subdued or buried crater rimcrest

Closed depression

Caldera

Lava flow lines

Knobs

Landing site target

Site 112 GUSEV CRATER

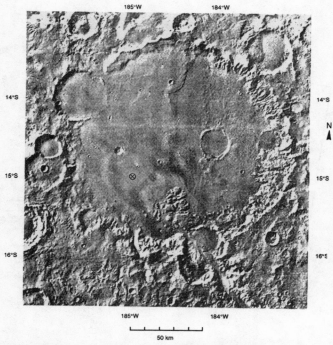

185°W   184°W

14°S   14°S

15°S   15°S

16°S   16°S

N

185°W   184°W

50 km

MTM -15182
MTM -15187

⊗ Landing Site: 15° S Latitude, 185° W Longitude

MESUR
(Ronald Greeley)

Site Name: Gusev Crater

Type of Site: MESUR

Latitude: 15° S
Longitude: 185° W
Elevation: +0.5 km

Maps: MC-23 NE, SE
      MTM -15182
      MTM -15187

Viking Orbiter Images: 434S08, 69 m/p;
  432S11, 70 m/p; 432S13, 70 m/p

Date Entered: April 1992
Date Last Revised: 8 March 1993

Contact:
Ronald Greeley
Department of Geology
Arizona State University
Tempe, AZ 85287-1404
(602) 965-7029

Geologic Setting

Landing ellipse falls on lacustrine deposits in Gusev Crater. This site is a potential ancient lake bed with deposits filling the crater floor. It has channels leading into and out of the crater floor.

Scientific Rationale

Lacustrine deposits (Hch).

Objectives

Obtain composition of lacustrine deposits, volatile information, exobiology, seismic information, meteorology, and photogeology. Site is one station of a proposed 16 station network to demonstrate the feasibility of a network configuration which meet the need of seismology, geology, geochemistry, and atmospheric science.

Potential Problems

To be determined.

Trafficability

Not applicable.

Estimated Traverse Distance

Not applicable.

This is just a taste of the possibilities. Many more sites have been identified, and more will follow the current detailed mapping. For instance, early in 1998 the thermal emission spectrometer aboard *Mars Global Surveyor* detected a huge deposit of hematite, more than 200 km in greatest extent. It could be a hydrothermal deposit, though it is much larger than any on Earth.

A problem that is often raised is that Mars is blanketed by a layer of broken rock and dust ejected from the numerous meteorite impact craters that pepper its surface, and therefore we will have trouble finding the rock below this regolith. The images that have been obtained show that this is not a great problem. Despite the fact that in places the regolith could be kilometres thick, in many places original structures like craters, valleys and rock layers are clearly and sharply exposed, showing that the problem, while real, is not a barrier to exploration.

What I and everyone else advocate for Mars is the 'picking winners' approach of going to a small number of sites that will be investigated in as much detail as possible. It is moot as to whether this will maximise the chance of success. In the discussions and reports that I am aware of, there is the tacit assumption that this is the only possibility given current technology and levels of funding. It is a high-risk strategy reliant on our powers of prediction. Palaeontologists have demonstrated that we can make this strategy work on Earth, but we are working from an enormous knowledge-base acquired over more than a century by tens of thousands of scientists on foot in the field, in laboratories, and using aircraft and satellite remote sensing. Maybe the pessimists who say we will never know whether there was life on Mars until we go there in person are in fact realists and we should cut straight to the main chance and put geologists on the planet.

IS THERE STILL LIFE ON MARS?

If there ever was life on Mars, there is a good chance it is still there. At least underground there will be liquid water, where the internal heat of the planet melts the permafrost. There is life on Earth in such environments, and there is no reason I can think of why it should not be so on Mars. As on Earth, the organisms in such an ecosystem could have a metabolism based on linked oxidation of hydrogen and reduction of carbon dioxide. Reaction of water with basalt produces hydrogen, and carbon dioxide is abundant in the atmosphere of Mars.

As outlined in chapter 1, on Mars the angle of the spin axis to the ecliptic (the plane of rotation around the Sun) has ranged from 13° to 47°, over the last few million years. When the obliquity is at a maximum, the polar caps would melt in the summer. The water and carbon dioxide stored at the poles would vaporise and be released into the atmosphere, possibly raising the atmospheric pressure high enough to make liquid water stable for short times. At such times any subsurface microbiota that might exist would migrate to the surface.

CHAPTER 7
MISSIONS TO MARS

We shall not cease from exploration and the end of all our exploring will be to arrive where we started and know the place for the first time.

T. S. Eliot, Little Gidding, *The Four Quartets*

By 1960 when the first mission to Mars was attempted, astronomers no longer considered that there were canals or any other evidence for advanced forms of life on the planet. But there was still scientific support for the possibility of Mars being vegetated. Not only had seasonal changes in colour been observed, but spectral observations suggested the presence of organic molecules. The possibility of life on Mars was very real for many people, and the search for it was central to the space program. 'The scientific question at stake in exobiology is, in the opinion of many, the most exciting, challenging, and profound issue, not only of this century but of the whole naturalistic movement that has characterised the history of western thought for three hundred years' (Space Science Board of the United States National Academy of Sciences, 1962).

Since those first missions in 1960, the exploration of Mars has included some of the greatest achievements of planetary exploration:

1962   Second spacecraft to fly by another planet—*Mars 1*, USSR (the
       first was *Venera 1* to Venus, also from the USSR)
1971   First spacecraft to orbit another planet—*Mariner 9*, USA
       First soft landing on another planet—*Mars 3*, USSR
1976   First Mars landing to return substantial data—*Viking 1*, USA
1997   First rover on another planet—*Pathfinder* and *Sojourner*, USA
       (though there was a manned rover on the Moon in 1971)

The 21-year gap from 1976 to 1997 reflects the doldrums of planetary exploration, as can also be seen in the list below. The magnitude of the exploration campaign can be gauged from the list of missions, 31 in all so far, a veritable armada. All except the latest are from the USA and the USSR (or Russia). This list is evidence of judgements about the significance of the exploration of Mars. The motives are scientific, especially the search for life elsewhere and understanding the evolution of another planet so we can better understand our own; and political—national prestige and dominance; and economic—support for local industry and capturing spin-off benefits. Also in the mix are those intangible benefits—leadership, motivation and inspiration.

Mission	Country/Launch date	Arrival date	Result
No name	USSR 10 Oct 1960	–	Failed at launch
No name	USSR 14 Oct 1960		Failed at launch
No name	USSR 24 Oct 1962	–	Failed in Earth orbit
Mars 1	USSR 1 Nov 1962	–	Lost communications
No name	USSR 4 Nov 1962	–	Failed in Earth orbit
Mariner 3	USA 5 Nov 1964	–	Shroud failed to detach
Mariner 4	USA 28 Nov 1964	14 July 1965	Flyby, first pictures of Mars
Zond 2	USSR 30 Nov 1964	–	Lost communications
Zond 3	USSR 18 July 1965	–	Lost communications
Mariner 6	USA 24 Feb 1969	31 July 1969	Flyby, many pictures of Mars
Mariner 7	USA 27 Mar 1969	5 Aug 1969	Many pictures of Mars

Mission	Country/Launch date	Arrival date	Result
No name	USSR 27 Mar 1969	–	Failed at launch
No name	USSR 14 Apr 1969	–	Failed at launch
Mariner 8	USA 8 May 1971	–	Failed at launch
Kosmos 419	USSR 10 May 1971	–	Failed at launch
Mars 2	USSR 19 May 1971	27 Nov 1971	Orbited 3 months, lander failed
Mars 3	USSR 28 May 1971	2 Dec 1971	Orbited 3 months, lander returned 20 secs of data from surface
Mariner 9	USA 30 May 1971	13 Nov 1971	Orbited and returned many pictures for first global maps of Mars
Mars 4	USSR 21 July 1973	8 Feb 1974	Failed to orbit but returned pictures
Mars 5	USSR 25 July 1973	12 Feb 1974	Orbited 10 days and returned 60 pictures
Mars 6	USSR 5 Aug 1973	12 Mar 1974	Lander crashed
Mars 7	USSR 9 Aug 1973	9 Mar 1974	Lander missed Mars by 808 miles
Viking 1	USA 20 Aug 1975	19 Jun 1976	Lander & orbiter. Landed 20 July 1976, operated for 5 years on Mars and over 4 years in orbit
Viking 2	USA 8 Sep 1975	7 Aug 1976	Lander & orbiter. Landed 3 Sept 1976, operated for nearly 4 years on Mars and 2 years in orbit
Phobos 1	USSR 7 July 1988	–	Command error turned antenna away from Earth, all contact was lost Sept, 1988
Phobos 2	USSR 12 July 1988	29 Jan 1989	Entered orbit around Mars, obtained images of Mars and *Phobos.* Contact was lost, 27 Mar 1989

Mission	Country/Launch date	Arrival date	Result
Mars Observer	USA 25 Sep 1992	25 Aug 1993	Contact lost just before orbital insertion
Mars Global Surveyor	USA 7 November 1996	12 Sept 1997	Orbiter
Mars 96	Russia Nov 1996	–	Failed soon after launch
Mars Pathfinder	USA 4 December 1996	4 July 1997	Lander and microrover
Nozomi	Japan 4 July 1998		Orbiter

Each mission represents a great engineering achievement. There also have been many scientific achievements. For instance, we now know the composition of the atmosphere of Mars, we have reconnaissance-level geologic and topographic maps of the whole planet, the two *Viking* landers have searched for life and in the process learned a lot about the chemistry of the surface of the planet, and most recently the *Sojourner* rover analysed the elemental composition of some Martian rocks, with surprising results (as I will discuss below).

After a decade with no American missions, a new era of exploration of Mars began with the launch of NASA's *Mars Observer* in 1992. Just as it was about to go into orbit around Mars in August 1993, contact was lost forever. From this failure which involved the loss of a mission costed at about a billion US dollars, and from earlier initiatives, has grown the Mars Surveyor Program of less ambitious, lower-cost missions. Included in the program is a series of missions to Mars involving two launches every 26 months, during the 'launch window'— a minimum-energy mission launch opportunity which occurs every 2.14 years, the time that elapses between Earth and Mars passing each other in orbit. This closest approach to each other is called opposition. Because the

125

orbit of Mars is eccentric, the distance between the planets at opposition varies by as much as a factor of two, affecting the energy required to launch a mission and the payloads that can be carried.

The American (and now also Japanese and European) focus on Mars will transform our knowledge of the planet within a decade. The early stages of the program are basic science and 'engineering demonstration' missions. The strategy is to develop the technology and gather the information needed for a major goal of the program, the first 'sample-return'. It is widely agreed that returning samples for analysis on Earth is an essential step both in understanding the planet and in the search for life. When I first became involved in NASA's Mars exploration program, in 1987, it was predicted that we would have to wait at least until 2010 for the first sample-return. Then as one problem after another beset NASA and the space program fell into public disrepute, that goal slipped to 'maybe 2020', a date made all the more likely by the loss of *Mars Observer*. Then the Mars Surveyor Program was established and optimism spread. But when the Administrator of NASA, Dan Goldin, announced in December 1995 that a goal of the organisation would be to launch a sample-return mission in 2005, the general reaction of informed scientists I spoke to was, 'too soon': the necessary background studies would not be completed in time. Then the announcement of 6 August 1996, about 'the meteorite', changed everything. Whatever the final outcome of the studies of meteorite ALH84001, the suggestion that it contains evidence of life on Mars has galvanised everyone into action (or has been used to galvanise us all, the cynical but possibly correct view). Now the tentative plan is for the first sample-return mission to be launched in 2003.

The pace of exploration will be determined by the level of public interest. No-one who is old enough can forget how President Kennedy used the threat of the domination of space by the Soviet Union to declare in 1962 that the United States would land men on the Moon by the end of the decade, thus accelerating the space program to new heights of achievement. That first footprint on the Moon was made by Neil Armstrong on 20 July 1969. Given the level of interest in Mars at present, and barring a disaster of some sort, there will soon be a sample return mission to Mars, and it will be the first of a series.

The Russian *Mars '96* mission was launched on 16 November 1996 from Baikonur Cosmodrome in Kazakhstan. Aboard was an array of instruments provided by twelve countries. These were to measure physical and chemical properties on the planet's surface, and were housed in two small landers and several penetrators. But because of the failure of the *Proton* launch vehicle's fourth-stage booster the spacecraft never escaped Earth orbit. Any debris which survived re-entry is believed to have crashed near the west coast of South America on the day of the launch. The next day it was predicted that more debris (later determined to be a booster rocket) would re-enter over Australia, resulting in emergency services going on alert. Fortunately the predictions were inaccurate and if there was any surviving debris it would have crashed into the Pacific Ocean off the coast of Chile.

Mars Global Surveyor was launched on 6 November 1996, and arrived at Mars ten months later, to go into orbit for a detailed mapping program. It has a 'thermal emission spectrometer' which can map the distribution of minerals with a resolution of 3 km. This means that patches of minerals more than 3 km wide can be detected and mapped. Also on board is a camera with a resolution of 250 m, and a high resolution camera which can see

127

objects only a few metres wide. Its laser altimeter has a vertical resolution of 2 m.

On 4 December 1996 *Mars Pathfinder* was launched from Cape Canaveral in Florida, after $3\frac{1}{2}$ years of development. Aboard was the tiny rover *Sojourner*, just 60 cm long. *Sojourner* was the name suggested by twelve-year-old Valerie Ambroise of Connecticut in a world-wide competition. *Sojourner* was the size of a child's toy car but was equipped with a camera, an α-proton X-ray spectrometer which could do elemental analyses of rocks and soil, and other devices. This mission provided the first close-up photographs of Mars taken for twenty years. When it ceased after three months as a result of equipment malfunction it had been a spectacular success, in every way. The 'direct entry' trajectory (with no orbiting), airbag landing (it bounced at least 15 times) and deployment of the rover (which travelled about 100 m altogether) all went well, and large numbers of new images (550 from the rover) and analyses (sixteen in total) are now available. Rocks at the landing site have proven to be more silica-rich than expected ('andesitic' instead of 'basaltic'), indicating that we have much yet to learn about the history of the planet. Another significant result is that the 'moment of inertia' of the planet was calculated and indicates that it has a central metallic core.

The amount of public interest in Mars was dramatically demonstrated by the number of 'hits' on the NASA web site after the *Pathfinder* landing. The peak was on 8 July, with 47 million! The July total was more than 400 million. NASA's policy of public openness and access and rapid provision of data is an effective way to ensure continued public interest and funding.

THE NEXT FEW YEARS

We are entering a period of international cooperation in space exploration, and the development of new and innovative technologies. There are now shared missions and new countries becoming involved. Japan's *Nozomi* is on the way, and *Mars Express* from the European Space Agency is due for launch in 2003. Since the loss of *Mars '96*, the Russian space agency has focused on its space station *Mir*, and the development of an international space station. They have devoted a small amount of money to building an instrument to be incorporated into the US *Mars Surveyor* lander. This will be the first time a Russian instrument has flown on an American mission. China shows no sign of active interest, as far as I know.

Amongst the new technologies is precision landing. Landing spacecraft in the intended spot is obviously desirable. *Pathfinder* landed only 23 km from the centre of its 300 by 100 km target ellipse. Anywhere within the ellipse would have been acceptable with current technology. The *Mars Surveyor* 2001 mission will attempt the first precision landing, aiming to be within 10 km of the designated landing site. This is expected to be possible because of improved interplanetary navigation and other advances.

In 2003 using an American *Delta* rocket NASA will launch a lander with a sample-collecting rover, a 'Mars ascent vehicle', and a drill and a robotic arm from the Italian space agency. Over several months the rover will collect samples and transfer them to the ascent vehicle which will then launch them into a low orbit. Then in 2005 a French *Ariane 5* rocket will launch a duplicate lander, rover and ascent vehicle, plus a French orbiter. This second rover will collect more samples from sites different from those of the 2003 mission, which

129

then also will be launched into orbit. Meanwhile the French orbiter will have been inserted into an elliptical orbit from where it will rendezvous and dock with the two sets of samples in their 'Earth entry capsules'. The orbiter will then fire its engines to propel itself back to Earth. As it gets close to home it will eject the entry capsules for landing back on Earth, and then deflect itself, so that it misses us. Landing sites for the capsules are currently being considered. There are favourable sites in the Southern Hemisphere, particularly Australia. The plans seem bland and matter-of-fact in NASA documents, but how amazingly audacious they are! It seems like a very fragile 'mission architecture' to me, with lots of scope for things to go wrong. If it works, the intention is to do the whole thing again several times.

Other proposals include sending giant balloons that will drift around the planet, rising through the atmosphere when they heat up during the day and settling on the surface at night, so carrying their instruments all over the planet. Orbiters will drop spear-like 'penetrators' into the surface from great heights, carrying instruments deep into the regolith. In the European Space Agency there is a strong emphasis on the need to drill as deeply as possible to get the most significant samples.

Mission Plans as of April 1999

Mission	Launch Date	Objectives
Mars Climate Orbiter	1999	Analysis of water, volatiles and climate
Geochemical Mapper	2001	Analysis of elemental composition and global minerology

130

Mission	Launch Date	Objectives
Mars Express (ESA/ASI)	2003	
Sample Return (NASA)		Sample collection
Sample Return Orbiter (CNES)	2005	Retrieval and return to Earth of '03 and '05 samples
Sample Return Orbiter (CNES)	2009	Retrieval and return to Earth of '07 and '09 samples
Sample Return Orbiter (CNES)	2013	Retrieval and return to Earth of '11 and '13 samples

BEYOND 2005

Inevitably someday there will be crewed (manned) missions to Mars. It is a challenge that is irresistible. It is frequently stated that such missions will not happen until well into next century because we should first work through a systematic program of observation and mapping from orbiters, unmanned rovers and other instruments. There are said to be recalcitrant technical problems to overcome, particularly health problems caused by zero gravity and cosmic radiation while in transit. But there is a growing band of articulate and informed specialists and enthusiasts advocating an early move to this phase of exploration. A former Associate Administrator of NASA wrote some years ago that the American people would not have supported prolonged robotic exploration of the Moon: the excitement generated by human missions was needed to keep the money flowing. Presumably the same will be true for Mars. Some dispute this logic, afraid that it will divert a huge amount of funding away from good planetary robotic science into premature human exploration, but they may be voices in the wilderness. Daniel Goldin, the

131

Administrator of NASA, stated in July 1997, at the height of the success of the *Pathfinder* mission, that he would support a crewed Mars mission if it would cost less than US$25 billion. A commitment by 2004 would allow such a mission as early as 2011. Presumably Goldin thinks it is possible. Those who argue against crewed missions are the space exploration equivalents of economic rationalists, or 'black letter' lawyers. We can call them planetary rationalists. They may have logic on their side, but life is not logical.

There is also a compelling argument that while robots can be made fairly smart, we are a great deal smarter and can achieve much more. Every observational, 'natural' scientist, such as geologists and ecologists, knows the value of years of experience. The natural world is enormously intricate and complicated. The ability to make intuitive decisions and creative choices between options, especially when faced with unexpected situations, is critical to good science. One day robots might gain some of these skills. We already have them. Albert Einstein wrote: 'The really valuable thing is intuition'.

Speculation by science fiction writers has played a significant role in keeping fresh and alive the possibility of people going to Mars. The eloquent and technically sophisticated stories by Kim Stanley Robinson and Ben Bova are outstanding recent examples. Is it just a dream? I don't think so. We can ask whether Mars seems any further away or any more of a technical challenge to us than 'Cathay' did to Columbus in 1492, the Pacific Ocean to Magellan in 1520, or Australia to Cook in 1770, or even, in our own century, than Antarctica did to the early explorers. Fifty years ago only dreamers believed that we would ever walk on the Moon.

In an extraordinary book, *The Case for Mars*, Robert Zubrin and Richard Wagner set out a rationale for colonising Mars. They attempt to demonstrate that it is technically feasible and economically achievable right now. They argue that there is no need to first establish bases on the Moon, and that the problems of spaceflight that are often raised, such as the physical effects of zero gravity and cosmic radiation, have been exaggerated or can be overcome. They favour mission trajectories involving the 'conjunction option' (launch when the planets are on opposite sides of the Sun), with six months travel each way and a stay of 550 days on the surface of Mars. This length of spaceflight has already been achieved by Russian cosmonauts aboard space station *Mir*, with apparently few ill effects (though they were shielded from cosmic radiation by the magnetic field of the Earth). Provided rocket fuel for the return journey is manufactured on Mars, that crews total only four people per mission, and grandiose 'Battlestar Galactica' style missions are avoided, then the missions can be launched with rockets such as the *Saturn 5* used in the Gemini program of the 1960s, or the Russian *Energia*. They argue that there is no need to assemble elaborate spacecraft in orbit around Earth, and no need for a space station. Zubrin and Wagner provide a detailed manual addressing key technical problems including the manufacture of the fuel. Both they and former astronaut Michael Collins in his book *Mission to Mars* see Mars as the next great challenge for human exploration. This is a challenge that inevitably we will accept, as we have accepted all other geographic challenges. As President Kennedy said of landing men on the Moon, this 'goal will serve to organise and measure the best of our energies and skills . . . This is in some measure an act of faith and vision, for we do not know what benefits await us . . .'

There are serious plans to establish bases on the Moon and Mars. Various types of habitats are being studied, including inflatable units. There are even scientists who

133

propose 'terraforming' Mars, that is, altering the surface environment of the whole planet to make it habitable again, as it was billions of years ago. Giant orbiting mirrors could focus the heat of the sun on the Martian polar caps to melt the dry ice and ice. By this means water and carbon dioxide can be put back into the atmosphere. Others propose using nuclear power plants for the same purpose.

Science fiction writers have long written about making Mars habitable, at least locally in enclosed bases or even cities. None of this can any longer be considered outrageously improbable. The pace of exploration is increasing and will lead inevitably to the establishment of exploration bases on Mars. There have already been fourteen successful missions to Mars, including four landers. Despite the gainsayers, there may soon be people on Mars, and the necessary research is already a major part of NASA's program, and a component of that of other space agencies. As mentioned in the Epilogue, there are the first stirrings of serious thoughts about commercial exploitation of space, including Mars.

ANTARCTICA

There is a close analogy between the human exploration of space and that of Antarctica, and for this reason there is a joint US–Australian project to study the physical and psychological health of Antarctic expeditioners. It concentrates on the Australian experience because this most resembles that of long-duration space flight. Unlike the US, Australia has no air service to its Antarctic bases, with the result that all expeditioners stay for at least the summer, and many winter-over in the extremely harsh environment. So the experience is comparable to the stresses and isolation of space.

Responses to the experience include a lowered immune function, and headaches, insomnia and gastrointestinal problems. Despite rigorous pre-expedition screening there have been serious medical and psychological problems. Each expedition has one medical doctor, backed up by a system of telemedicine. Appendixes have had to be removed, and even more serious operations performed, with the help of non-medical expedition staff.

A few examples from the Australian Antarctic experience help us envisage the situations that can occur. For instance, pity the poor doctor who himself has a medical problem: Dr Peter Gormly once had to drill his own tooth abscess while looking in a mirror. Or consider what can happen in a real emergency: in 1961 mechanic Alan Newman suffered a potentially fatal brain haemorrhage. The doctor, Russel Pardoe, had no experience of neurosurgery and lacked the vital instrument. After consulting via radio-telegram with a neurosurgeon back in Australia, and with the aid of illustrations in an instrument catalogue, a brain cannula (tube) was improvised. Then the operating procedure was tested on a seal that had been shot in the head with a .38 revolver. The first operation on Newman was partially succesful, but had to be repeated a few days later. The expedition cook acted as theatre sister, and the geophysicist as assistant anaesthetist. Despite this heroic effort Newman continued to deteriorate. Fortuitously and very fortunately he was able to be evacuated by a Soviet aircraft and operated on again in Sydney. He made a complete recovery.

Each expedition group develops its own culture, in ways that have proven unpredictable, and there are many reports of bizarre behaviour. A cook fed up with one man's carping complaints about the occasional hair in his food baked him a hair pie. Ritual bathing in freezing waters seems to be *de rigueur*. There have been no murders at Australian bases, though what looked like a

135

serious attempt was averted once. Before a program of psychological screening was introduced in the 1960s one man became deranged and had to be isolated in a makeshift cell for six months. The rarity of any substantial problems now is a tribute to both the screening and to human nature.

For those of us who will be mere observers of human travel to Mars it is amusing to contemplate the composition of the crews. It will be a deadly serious issue for those involved. Finding volunteers will not be a problem, as I have learned from talking to many people. Imagine, though, a Zubrin and Wagner style crew: just four people, cooped up together for six months' travel each way plus 550 days on Mars. Try this party game at the dinner table some night: should it be two married couples (of whatever combination of genders), or maybe all women (because they eat less)? Should there be no medical doctor (as advocated by Zubrin and Wagner) because they cause people to focus on problems, and conduct intrusive tests? How many should be scientists and how many technicians and pilot-type astronauts? They will have to be multi-skilled. Is it easier to train a pilot to be a geologist, or vice versa? Should they all be young, or will the wisdom of age be important? Prepare yourself for some good arguments.

CONTAMINATION

Public concern about the contamination of Mars by Earth organisms, or vice versa, could become a major constraint on the exploration program. The same issue was important during Lunar exploration, particularly when samples were returned. Extraordinary precautions were taken. The procedures for sterilising spacecraft and for isolating returned samples are well established. A story often told is that viable bacteria were found in a camera returned to Earth after two years on the Moon. In other words, we could not rely on the extremely hostile surface conditions on Mars to kill any organisms that escaped the equipment sterilisation procedures on Earth and the hostile conditions on the journey to Mars. However, it is considered that this example is not valid, and the bacteria were introduced during a failure of experimental protocol after the camera was returned. Whatever the correct interpretation of this incident, the fact is that the issues of 'forward contamination' and 'back contamination' are taken very seriously. But in the end no-one can deny that there is still some risk. We each must decide for ourselves whether the benefits of exploration outweigh the risks.

'Forward contamination' is the degrading of another object in the Solar System by microbes or chemicals from Earth. This can be treated as an ethical issue analogous to the protection of wilderness areas on Earth. It is also of pragmatic concern: contamination could reduce our opportunities for contrasting another planet with our own. Even more directly, if we contaminated samples with microbes from Earth and then detected those microbes in the receiving laboratory on Earth, we might decide not to risk releasing the samples for research because the microbes might be alien. Spacecraft are sterilised to minimise these risks. Techniques include the use of hydrogen peroxide gas plasma, gamma radiation, ultraviolet radiation and heat. It has been argued that the surface of Mars is so inhospitable that no Earthly microbe could survive there. That may be correct, but at our current level of knowledge of Mars it is not a convincing argument.

Some scientists insist that Earth and Mars have never been quarantined from each other anyway, having participated in mutual meteoritic trafficking throughout their history. However, transfer of microbes by this mechanism

137

is a prediction, not an observation, and is something we would wish to confirm or refute by making observations on a Mars free of human contamination. Once crewed missions commence, the problem will be even greater, because we carry with us in and on our bodies an abundant and diverse microbiota (half the cells in our bodies are microbes).

Much more difficult than killing microbes is cleaning spacecraft so that there is no contamination by organic compounds from Earth. The detection and analysis of organic compounds on other bodies in the Solar System is a major goal of many exploration programs, because of the significance of this for understanding early steps in the origin of life, and a dirty spacecraft would compromise these efforts.

'Backward contamination' is concerned with returning alien microbes to the Earth, with all the risks that might entail. As many missions will be seeking evidence of life elsewhere, the plans include isolating samples in sealed containers that can survive re-entry to Earth, and then testing them in a receiving laboratory constructed to the highest standards of biological containment. Such a laboratory was used for the samples returned from the Moon, and there are comparable laboratories in constant use for the study of highly dangerous bacteria and viruses. So the technology exists and is demonstrably effective. No samples would be released until they had been tested exhaustively for any deleterious properties, and then only to organisations with appropriate facilities. It is said that even if Martian microbes were returned to Earth they would not be pathogenic because they would have evolved out of contact with all Earthly organisms, and therefore would not be chemically compatible. Given the propensity of microbes to participate in lateral transfer of genes, I do not find that reasoning very comforting. Of course if the analyses were done on Mars or the Moon there would be much less concern.

These issues of 'planetary protection' are covered by the Outer Space Treaty of 1967. A UNESCO commission has an advisory role. The risks are taken very seriously, but as more nations explore space, it is not clear that they all adhere to the same high standards. It may be invidious to single out any one nation, but it was remarkable when at a recent international space exploration conference in Japan, in July 1998, the NASA Planetary Protection Officer, John Rummel, was asked whether he was aware of what measures had been taken to deal with contamination risk from the Japanese *Nozomi* mission to Mars, he could not answer. There apparently has been no international consultation. Japanese scientists in the audience did not comment.

A similar telling event occurred during the 49th International Astronautical Congress in Melbourne in September 1998. On the stage of the main auditorium in the convention centre were E.C. Stone, Director of NASA's Jet Propulsion Laboratory; P. Wenzel, Head of the Solar System Division of the European Space Agency; A. Nishida, Director General of the Japanese Institute of Space and Astronautical Science; and I. Galeev, Director of the Institute of Space Research, Russia. Each of these people made presentations about their agency's plans for exploration of the Solar System, and then the group as a whole took questions from the audience of several hundred scientists. Gerald Soffen, chief scientist for NASA's *Viking* missions and a very senior scientist within NASA asked what was being done to prevent forward contamination of the objects to be visited in forthcoming missions. Only Ed Stone replied, and his answer was bland and restricted to NASA's responsibilities. The other three declined the opportunity to comment.

The issue of contamination is certainly not being ignored, but the suspicion is that it is so difficult and expensive to deal with thoroughly that short-cuts are being taken, by all space agencies. My own view is that the risks should be more widely known and acknowledged so there can be informed public debate.

PRAGMATISM AND INSPIRATION

The current NASA missions to Mars cost the people of the United States about a dollar each per year. Cost overruns on the development of the International Space Station are inhibiting the Mars program. This is officially denied, but Solar System exploration projects have been cut back recently, and many scientists involved certainly believe that funds are being diverted from their projects to the space station.

We all hear about spin-off benefits of space exploration but, as written by a United States government committee, 'perhaps the most important space benefit of all is intangible—the uplifting of spirits and human pride in response to truly great accomplishments'.

In the absence of a great dream—pettiness prevails

Robert Fritz

CHAPTER 8
SO WHAT?

On talkback radio shows I have been told that spending money on this esoteric nonsense is an obscenity. People are starving. There are wars. Disease is ravaging the Earth. We are destroying the environment. Most people, though, find inspiration and excitement in space exploration and see value in the knowledge gained, not to mention the industrial and economic benefits.

Could Earth be the only inhabited place in the Universe, perhaps because God made it so, as many people believe? Or maybe the steps leading to the origin of life are just so improbable that they happened only once? These seem to be questions of significance only to those of us in the Western tradition. For instance, a Hindu colleague told me that of course there is life everywhere: Hindus have always known that to be so.

People have long searched for evidence of life elsewhere, but the modern version of the Search for Extraterrestrial Intelligence (SETI) program began in 1960 when American astronomer Frank Drake used a radio telescope in the United States to listen for signals from two stars. Since then 85 other programs (most using radio telescopes) have searched for evidence of technologically advanced civilisations around other stars. Today four such programs are continuing, three based in the United States and one in Argentina. The most

141

comprehensive of these is Project Phoenix. Like the mythical bird, this project rose from the ashes, in this example from one originally funded by NASA but eliminated by Congress. It is now funded entirely by private donations. Phoenix uses equipment that can be deployed at various telescopes and during 1996 was based for six months in Australia. By the end of this century the search will have covered about 1000 stars within 150 light years of Earth. Needless to say, none of the searches has been successful, otherwise you would have heard. Even one of the strongest advocates of this program, Dr Jill Tarter of the SETI Institute in California, said in a speech to the Pontifical Academy of Sciences in 1996: 'The searches may need to extend over generations before they are successful'. The reasons for this lie in the probability of there being intelligent, technological life elsewhere, as explained below.

No telescope is powerful enough to see planets around other stars. But astronomical observations over just the last two years have discovered more than nine planetary systems beyond our own. These systems have been discovered by observing the systematic wobbles introduced in the rotation of stars by the presence of orbiting planets. The technique relies on detecting shifts in the frequency of light emitted by the stars (the 'doppler shift'). At present it is sensitive enough only to detect the effects of giant planets with masses comparable to that of Jupiter. The reason that these discoveries have all occurred recently is that the necessary instrumentation was first built in 1994. Observations must be made over periods of months or years to detect the orbital wobbles, so it will take a long time to build up a clear understanding of the abundance of planetary systems. Another discovery was made by measuring minute changes in the rate of rotation of a pulsar, a collapsed, neutron star that emits radio waves and spins rapidly so that the radio emissions sweep across space like a beam from a lighthouse. Just recently another possible planet was detected using a different technique, 'gravitational lensing'. In this approach, light from a distant star is focused by the gravity of another star between it and Earth. If the second star has an orbiting planet then that also has a detectable effect. This most recent observation has yet to be confirmed. Yet another recent discovery is of dust clouds around many stars; this is the material from which planets will form. In the light of all these discoveries, any suggestion that Earth is unique has become increasingly untenable.

In 1961 Frank Drake tried to estimate how many detectable extraterrestrial civilisations there might be, using a now-famous equation to assess all the relevant probabilities. This includes terms for the probability of there being other galaxies and planetary systems, other Earth-like planets, life originating elsewhere, and so on. His most recent published estimate is that there could be about 10 000 other civilisations in our galaxy alone. Australian physicist and philosopher Paul Davies considers that this underestimates the great improbability of the origin of life. Davies' point is that theorists have not taken sufficient account of the difficulty of generating the huge amount of information encoded in even the simplest form of life; even the simpler precursors of the first single cells would have included numerous 'information-rich' molecules. When this is considered, and in the absence of as yet undiscovered principles, the origin of life seems less probable than ever. No scientist doubts that the various probabilities involved are extremely poorly known. So there might be only ten other civilisations in our galaxy, or only one. No one knows. As Davies himself has reasoned in his book *Are We Alone?*, the Universe is, in a human perspective, infinitely large. There is an extremely large number of

143

galaxies. So when we attempt to calculate the probability of there being intelligent civilisations elsewhere, even if one term such as the probability of the origin of life in the Drake equation is very small, another term like the number of galaxies can be extremely large. Multiplying all of the terms together might suggest that there is a significant probability of there being intelligent life elsewhere. But the nearest other intelligent community might be so far away as to be outside the observable Universe. Perhaps none of this is very meaningful because the probabilities involved cannot yet be convincingly evaluated, but it does encourage us to keep looking.

Conspiracy theories demand a brief mention. There are theories that SETI has already found evidence of life, or that NASA already knows there is life on Mars, and 'the government' has hushed up the news. Anyone who harbours such notions should remind themselves how bad governments are at keeping secrets; someone always leaks the information, eventually. Scientists are even worse with secrets: we can't wait to tell everyone when we make a discovery. There would be no chance of keeping the discovery of life elsewhere secret. And why would anyone want to, anyway? To prevent panic, we are sometimes told. But when many or even most people already believe there is life elsewhere, why would there be panic? Even if you think governments can keep secrets, in the example of Mars consider these three facts: there is no magnetosphere to deflect lethal cosmic radiation from the Sun, there is no ozone shield against lethal ultraviolet radiation, also from the Sun, and the surface is highly oxidising, which is also a fatal condition. All the evidence is that these conditions have existed for at least millions and probably billions of years. The possibility of life unprotected on the surface fashioning giant faces and building pyramids is vanishingly small.

MICROBIAL LIFE

The Drake equation assesses the probability of there being advanced civilisations with 'technological intelligence'. We need to consider the relationship between 'technological intelligence' and 'life'. Look back at the universal tree of life in chapter 2 and consider the occurrence of intelligence. The total number of species of all organisms on Earth is not known except that it is more than four or five million. Some recent estimates go as high as 100 million. Biologists have a very long way to go in documenting the diversity of small and inconspicuous organisms of many kinds, especially microbes, so it could be even greater than this. A complete tree would have millions of twigs, one for each species. And those are just the species alive now. Many millions of others are long since extinct. One twig can be labelled 'intelligent'. We can extend our concept of intelligence to include dolphins, whales and chimpanzees, if we wish. But even if we choose to include chimps in our calculation because they differ from us in only 3 per cent of their genes, it does not change the result. If we define intelligence as including the sort of technological ability that would be detectable in the SETI programs, that is, the generation of radio signals that we could detect with our radio telescopes, we are back to one twig. And that twig is very young, as *Homo sapiens* evolved less than half a million years ago (maybe only 100 000 years ago), in a tree that is 4 billion years old.

Then consider this: what we call civilisation has a history extending back 5000 to 10 000 years, but radio was invented less than a century ago. So a SETI program based in another planetary system would not have been able to detect us until this century. This form of technological intelligence has existed on Earth for no more than 0.000000025 per cent of biological history. How long will such intelligence last? Given our destructive ability it might not be long. Even if we do survive for more than the few million years normal for any species, a meteorite or comet might get us in the end. It is estimated (controversially) that giant meteorites or comets strike Earth every 20 million years or so. Some, at least, of these seem to have caused massive extinctions on a global scale, the best known of which killed off the last of the dinosaurs 65 million years ago. Our species could survive for millions of years, which from our perspective seems like an infinity, but even then our time is unlikely to exceed 0.25 per cent of the full span of biological history. (I have been optimistic and given us 10 million years, out of a total span of 4000 million.)

Other extinction events were even more devastating for life, such as that at the end of the Permian period, 253 million years ago. The cause of that particular event is not known, but palaeontologists frequently write that 95 per cent of species were wiped out at that time. What they actually mean is that 95 per cent of plant and animal species became extinct. As best we know, the microbial world was little affected. That is demonstrably so, because the fossil record shows that the main branches of the universal tree of life found by studying modern microbes have a very long history. If a large proportion of all microbial species had been lost in any extinction event, the tree would be dominated by recently evolved branches.

Microbes dominate life. Of the three superkingdoms of life recognised currently, Bacteria, Archaea and Eucarya, all of the first two and most of the third is microbial. Most of the chemical mechanisms that characterise all life, the internal cell machinery, no matter how 'advanced' or how big that life is, evolved within the microbial world and were inherited by the larger organisms more familiar to us. Microbes are not something apart from 'real' life: they are most of life on Earth, and include our ancestors. For the first two billion years or more of biological history on Earth, microbes were all that there was. Even now, the total weight of all microbes on Earth may equal the weight of all other organisms, based on predictions of the abundance of subsurface microbes by Cornell University scientist Thomas Gold and other recent workers. Microbes have dominated Earth for 4 billion years.

This reasoning suggests that for every example of an intelligent, technologically advanced civilisation, there will be thousands or millions of other planets inhabited only by microbes or the like. From this perspective, finding a fossil bacterium on Mars could be sufficient to establish with a high probability that we are not alone in the Universe. It would show that evolution of bacteria happened at least twice in our Solar System, increasing the probability that it would have happened on all the habitable planets in the Universe, of which there is likely to be a very large number. Nor would there be any reason to think that on many of those planets evolution would have stopped at the level of microbes. Evolving to complex organisms is likely to be a natural consequence of the competition for resources.

It is necessary to be cautious and write 'finding a fossil bacterium on Mars could be sufficient to demonstrate that . . . we are not alone', because one further step in logic is needed: we would have to demonstrate that any life found on Mars had an origin independent of that on Earth and did not, for instance, arrive there aboard a meteorite from here. If we have only fossils to work with, making that distinction is likely to prove extremely difficult.

I have glossed over a major theoretical problem: whether evolutionary events are random or directional.

147

We need to know this to be able to estimate the likelihood of microbial worlds elsewhere ultimately giving rise to intelligent beings. Many biologists would agree with Stephen J. Gould when he says that the whole process of evolution is based upon an infinite number of 'contingent' events. That is, that each new species evolves when a local environmental opportunity exists, and others become extinct for any one of many causes. Gould says that if we were able to 'rewind the tape of life and replay it' we would get a different result every time. Not only would the chance of getting humans a second time be nil, but it is very likely that a second rerun would not even produce any vertebrate animals at all—no fish, reptiles, birds or mammals, or any other familiar group of organisms. 'We are glorious accidents of an unpredictable process with no drive to complexity . . .'

A diametrically opposing view is put by Nobel Laureate Belgian biologist Christian de Duve in his book *Vital Dust.* He argues that chance plays a role, but within limits set by the physical and chemical properties of life. Evolution looks random when the whole range of species is viewed, but trim the tree of life 'of this outer diversity and you are left with a stark trunk delineated by a relatively small number of major forks . . .' While there is plenty of scope for evolution on another planet to have taken different pathways, 'certain directions may carry such decisive selective advantages as to have a high probability of occurring elsewhere as well'. For instance, 'Let something like a neuron once emerge, and neuronal networks of increasing complexity are almost bound to arise' because of the advantages they confer. This is a 'drive' in the direction of evolving larger brains, and, in the end, intelligence. That intelligence, though, need not inhabit a body at all like ours. 'Conscious thought belongs to the cosmological picture, not as some freak epiphenomenon peculiar to our own biosphere, but as a fundamental manifestation of matter . . . Thought is generated and supported by life, which is itself generated and supported by the rest of the cosmos.'

There is no way I can do justice here to these two profoundly opposed views of life. They illustrate how two eminent scientists working with the same facts can come to very different conclusions (there are intermediate points of view, but I have presented the extremes to make my point). The main reason such differences of opinion can happen is that we know only one example of a tree of life. If we had others we would not have to argue in the abstract, and could directly analyse the points of similarity between them, and from that deduce general principles of evolution. It is very hard to generalise from one example, and therefore to make predictions. What this does, of course, is provide us with a powerful scientific argument for looking for life elsewhere. Even if we had only two examples we would have a much better chance of determining the underlying principles of the origin and evolution, and meaning, of life.

PARADIGM SHIFTS

The usual way to think about the consequences of any discovery of life elsewhere is to compare such a discovery with other profound 'paradigm shifts'. For instance, the sixteenth-century discovery that Earth is not at the centre of the Universe, and the development last century of the theory of evolution. Both these scientific discoveries were seen as profound challenges to established social structure, and so were resisted by the full might of the Western churches. Many Christians still see evolution as a threat. It has challenged thinking about the meaning of life, by providing a natural explanation for the origin

149

of our own species. It is one of those profoundly significant ideas that cannot be ignored. It must be accepted or refuted, and either way has consequences in every walk of life.

Of course there are straightforward physical consequences of the theory of evolution. Darwin could not have foreseen the current debate about genetic engineering, or the possibilities raised by the birth of the cloned sheep 'Dolly'. In a little more than a century the understanding of evolution has influenced our lives in lots of practical ways such as the use of genetic engineering to make more productive and disease-resistant crops, bigger pigs, and a multitude of other examples, some negative. Who knows what practical consequences might follow from the discovery of life beyond Earth?

It is easier to imagine the philosophical consequences of such a discovery. For instance, Steven Dick, astronomer and historian at the United States Naval Observatory in Washington, considers that

> The question at stake in the extraterrestrial life debate is whether a . . . 'biological law' reigns throughout the universe, whether Darwinian natural selection is a universal phenomenon rather than simply a terrestrial one, whether there are other biologies, histories, religions and philosophies beyond the Earth. Put another way, we are trying to determine whether the ultimate outcome of cosmic evolution is merely planets, stars and galaxies, or life, mind and intelligence . . . More than fossils are at stake . . .

> Steven Dick, 'Humanity and extraterrestrial life' in *Life on Mars, What are the Implications*, Space Policy Institute, Washington DC, p. 34

The issue is the establishment of a new 'world view', a redefinition of our place in the Universe. Just as astronomers showed that Earth is not special in the sense of being at the centre of the Universe, astrobiologists could show that humankind is not special in the sense of being the pinnacle and goal of evolution. Most scientists might already accept that as a fact, but people at large have yet to be convinced, I believe.

More specifically, the possibility of life elsewhere raises a difficult problem for Christians. If Christ died to redeem the inhabitants of Earth, what of the inhabitants of other worlds? This is an old question, but still a fundamental one. It was one of the issues debated during 1996 at what, for me, was a very remarkable conference. The theme was the origin and early evolution of life. There is nothing special about that. What was special was that the conference was in the Vatican, and organised by the Pontifical Academy of Science. The participants included theologians as well as scientists. It was a week made even more remarkable by the release of a statement by Pope John Paul II acknowledging that most scientists accept the reality of evolution, and that this involves no conflict with Christianity. So what is the answer to the question of redemption? Teilhard de Chardin, in his book *Christianity and Evolution*, was troubled by this question. 'All that I can entertain is the possibility of a multi-aspect redemption which would be realized, as one and the same redemption, on all the stars . . . Yet the worlds do not coincide in time ! There were worlds before our own, there will be other worlds after it.' Not much of an answer, but the best I know.

My impression is that most people believe there is some form of life elsewhere. But being confronted with proof of such belief will still be a profound event. Any discovery of life beyond Earth will change us all forever.

WHAT WE STAND TO LEARN

The exploration of Mars presents us with five possibilities:

- If Mars is sterile now and always was, which would be very difficult, but perhaps not impossible, to prove, our conclusion might be bleak. A habitable place was never inhabited. Of course that would prove nothing about life anywhere else in the Universe. But as best we know at present Mars was the most likely habitat for life beyond Earth, so we would have to confront the possibility that we might be alone. We would have evidence that the origin of life is just so improbable that it happened only once. That might be comforting to those with vivid imaginations who fear star wars, or to anyone with spiritual beliefs requiring Earth to be special, but for others, I think, it would be a fearful vision. If any good were to come from this it would be an intense understanding of the fragility of life on Earth.

- If complex organic molecules were found, different from those in interstellar space and in comets and meteorites, we might conclude that life almost got started but was nipped in the bud by decreasing temperatures as the planet froze. We would have available for study a molecular museum which might include a Rosetta Stone which could show us critical steps in the origin of life. We might have preserved the first replicating molecules and the earliest membranes. This pre-life record has long since been destroyed on Earth, by frequent reworking of the early crust.

- If convincing fossils or extant microbes were found, and were able to be accommodated within the universal tree of life already known from Earth, then we would have to confront the reality of the transfer

152

of life by meteors. We would not know where life began, whether on Mars, on Earth, or somewhere else. Our exploration for both life's beginnings and its distribution would have to encompass much of the Solar System. For a few scientists, this would be no surprise, as they have already concluded that the spread of life around the Solar System is inevitable. The alternative, that life originated outside our Solar System and somehow made its way here, is not credible, though again, there are those who disagree.

• If the life discovered were familiar to us, but different in subtle but profound ways such as the chirality ('handedness') of its molecules, then we would contemplate the forces that guide evolution in particular directions. The power of self-organising systems would be manifest.

• If convincing fossils or extant microbes were found, and proven to be significantly different from any on Earth, we would conclude that life originated more than once. We could reason then that the Universe teems with life, and that it is just a matter of time before we detect it outside our Solar System, in the spectral signatures of other planetary atmospheres, or as radio signals. We can foresee such events, but what would follow, I think, transcends our imaginations. Communication seems fanciful, given the distances to even the nearest stars and the ultimate constraint imposed by the speed of light, but who dares say anything is impossible? Many of us have grandparents, or parents, who scoffed at the idea of men standing on the Moon.

Whatever proves to be the situation on Mars, we will learn a great deal about our own Solar System, and about ourselves. As we learn more about our sister planet, we will certainly look at Earth with new eyes. And we have a real chance to find answers to profound questions that fascinate us all. Even if there is not, and never was, any life on Mars, or anywhere else, the quest for it is a great intellectual endeavour. In attempting to discover this particular truth, we confront the definition of 'life', life's origin, its physical limits and its meaning. It is a quest that cannot fail.

The first people on Mars will find the sky pink from suspended bright red dust. The sunrises and sunsets may be even more beautiful than ours, though they might resemble those that follow wildfires in the Australian bush. Just before dawn on some nights there will be water-ice clouds, high in the sky; these will dissipate shortly after sunrise. The ground will be dry and barren, and strewn with grey rocks. Instead of soil there will be red dust. On many days, light winds will ripple the ground, and during the northern winter, great storms may develop, enveloping the planet in dust. In summer, temperatures near the equator will rise as high as 17°C during the day, but at night will plummet to minus 80°, or less. Perhaps somewhere, on the flank of a volcano, or deep in Valles Marineris, our explorers will find a place where the ground is damp, maybe in a fresh landslip, in the heat of summer. On that summer's day, in that Martian spring, they might just find the microbes that will show that Earth is not the only inhabited planet. And you and I could still be alive to contemplate that moment.

EPILOGUE

A small survey of Americans interested in space exploration found that 80 per cent considered it 'very important' or 'somewhat important' that 'the United States be the first nation to establish a continuing presence, either human or robotic, on Mars' (Space Policy Institute, George Washington University, 1997).

The rest of us should ponder the significance of that survey—who will control the future of Mars, and the rest of the Solar System?

During the 49th International Astronautical Congress in Melbourne in September 1998, James William Benson, Chief Executive Officer of the US corporation SpaceDev, delivered a paper entitled: 'Space resources: first come first served'. At first I thought it was just interesting speculation. Then I realised that he is serious and already has 70 aerospace engineers working for him. His plans include a robotic mission to an asteroid, during which he will claim ownership. He says there are no applicable laws.

It is in the best interest of humanity to quickly establish strong property rights in space. Investors need to know that their risk money can be rewarded through the exploration, discovery, ownership and utilization of abundant and easily reachable concentrated resources in space . . .

155

Today it costs less than $50 million (US) to fly a sophisticated deep space mission.

. . . in 1255 two Venetian brothers, the jewel-merchants Niccolo and Maffeo Polo, set out on the first European trading journey to China. After crossing the desert and the steppe, they entered Cathay itself and reached the imperial city of Cambaluc . . . They were well received by Kublai Khan . . . Their first trip occupied fourteen years yet just two years later they embarked again, this time taking not scholars and ambassadors but Niccolo's son Marco, a youth of seventeen whose subsequent experiences and recollections form the all-important opening chapter in the European exploration of the world (Peter Whitfield in his book *New Found Lands*).

The current program of exploration of the Solar System includes many examples of international cooperation. But when it comes to the first footprint, the first flag on a pole, even ownership, what will happen? How much of it matters, anyway? Will it be like Antarctica, where an international treaty protects the continent to some extent, or like most of the Americas, or Australia, where European will has prevailed? Should we think of space as a wilderness to be protected? We need to be prepared to answer these questions.

The example of Antarctica is instructive: many nations now have a say in the future of the continent. Not all are nations with great power. Australia, for example, has influence that flows ultimately from the pioneering exploration by a handful of scientists early this century, of whom geologist Sir Douglas Mawson is the most famous. As a result of international interest and concern, commercial exploitation is minimal and controlled.

We are at the dawn of a new age of exploration, which might turn out to be no less revolutionary than the classic

156

period of European exploration, the sixteenth to nineteenth centuries. 'The Europeans' true discovery was that all this knowledge could be merged into an accurate map of the world, which in turn became a vital tool of political power . . . Historically, this explosion of knowledge must be seen in the context of the intellectual revolution which we call the Renaissance, but the immediate motives of the explorers were overwhelmingly worldly—rapacious, mercenary, military and imperial' (Peter Whitfield again). Tough stuff, and incomplete to the point of being cynical, at least for some of the later explorers, but let's not forget that famous admonition, 'those who ignore history are doomed to repeat it'. 'Whatever the ideals with which explorers [of Africa this time] set out, ideals of serving geographical science or obeying a religious imperative, it was impossible that their discoveries would not be used to further European economic aims and political rivalries'.

Small nations like mine (Australia) cannot expect to be major players in space exploration. But as with Antarctic exploration, a small investment could produce big benefits. Perhaps the major benefit would be the one mentioned above—we would be taken seriously in discussions about future exploration and development. There are other benefits of space exploration, as everyone has been told for decades—microelectronics, teflon, on and on. Most significantly though, planetary exploration is inspirational, uplifting and exciting. That alone justifies the effort, and the money. That's my opinion, anyway.

> . . . it is necessary not to abandon the passion for ultimate truth, the eagerness to search for it or the audacity to forge new paths in the search . . .
>
> Pope John Paul II, 1998

GLOSSARY

Archaea Recently recognised superkingdom of organisms, all single-celled and lacking nuclei, previously classified within the Bacteria, but now known to be genetically and biochemically distinctive.

Astrobiology The study of life throughout the Universe, its origin, evolution, ecology and destiny.

Autotroph/autotrophy Organism able to make its cell material from inorganic sources of carbon, i.e., from carbon dioxide and dissolved bicarbonate ions, and the process of doing so.

Bacteria Superkingdom of organisms, all single-celled and lacking nuclei, distinct from Archaea and Eucarya.

Biogenicity The extent to which living organisms are or were involved in the origin of something.

Biomarkers Chemical remnants of cells found preserved in rocks, 'chemical fossils'. The term usually, but not always, refers to hydrocarbons derived from the components of cells.

Black smokers High temperature springs of water on the ocean floor, which appear black because of suspended particles of iron sulfide.

Chert   Rock composed of precipitated silicon dioxide, now quartz. A common example is flint.

Chirality   The property of some molecules to occur in two or more geometric forms which are mirror images of each other.

Cyanobacteria   Photosynthetic bacteria which use the hydrogen in water to reduce inorganic carbon to make cell material, so releasing oxygen. Previously called 'blue-green algae'.

DNA   Deoxyribonucleic acid, the molecule that stores the genetic information, genes, in cells.

Ecliptic   The plane of rotation of a planet around the Sun.

Enantiomers   The different forms of chiral molecules.

Eucarya   Superkingdom of organisms characterised, amongst other things, by cells having nuclei (in contrast to the Bacteria and Archaea); all the familiar large forms of life, as well as many microscopic forms.

Exobiology   The study of life beyond Earth.

Genome   The totality of all genes in any organism.

Heterotroph/heterotrophy   Organism unable to make its cell material from inorganic sources of carbon, and so requiring sources of organic molecules, and the processes involved.

Human Genome Project   An international project to map the entire genome of *Homo sapiens*.

Hydrothermal system   A natural hot-water system.

Hyperthermophiles   Organisms that grow best at more than 80°C.

159

Igneous   Previously molten rocks that have crystallised (e.g., basalt and granite).

Isotopes   Different forms of an element distinguished by the number of neutrons in the nucleus.

Lateral gene transfer   Natural processes by which some of the genes of an organism can be transferred into the genome of another species.

Launch window   The time opportunity for launching a spacecraft to another body in the Solar System involving minimal energy or the shortest flight time.

Microbiota   A population of microbes.

Microfossils   Fossil microbes.

Mitochondria   Small bodies within the cells of eucaryotes which are the site of oxidative respiration, i.e., where organic compounds are oxidised to produce energy.

NASA   National Aeronautics and Space Administration of the United States of America.

Obliquity   The angle by which the spin axis of a planet to the plane of its ecliptic differs from 90°.

PAHs   Polycyclic aromatic hydrocarbons, i.e., hydrocarbons in which the carbon atoms are arranged in rings which are themselves linked together.

Palaeobiology   The study of fossilised life.

Palaeontology   As for palaeobiology, except often considered to have more specific aims, for instance the classification of fossils and the dating of rocks.

Permafrost   Underground permanent or semi-permanent water ice, permeating soil and rocks.

Planetesimals   Rocky bodies the size of mountains or asteroids, which are considered to have been the building blocks of the planets.

Plastids   Small bodies within the cells of eucaryotes which are the sites of photosynthesis or the storage of starch.

Prebiotic   Before life, usually referring to organic chemistry.

Procaryotes   Archaea and Bacteria, in contrast to eucaryotes or Eucarya.

Regolith   The mantle of unconsolidated fragmental material that covers a land surface; i.e., soil and fractured rock.

Ribosomes   The molecular machines that synthesise proteins in cells.

RNA   Ribonucleic acid, one of the information storage and catalytic molecules in cells.

SETI   Search for ExtraTerrestrial Intelligence.

Sol   A Martian day, 24 hours and 37 minutes.

Stromatolite   Sediment shaped by the activities of mat-like masses of microbes.

FURTHER READING

Henry S.F. Cooper (1976). *The Search for Life on Mars: Evolution of an Idea*. Holt, Rinehart and Winston, New York, 254pp.

Paul Davies (1995). *Are We Alone?* Penguin Books, London, 109pp.

Paul Davies (1998). *The Fifth Miracle*. Allen Lane. The Penguin Press, London, 260pp.

Christian de Duve (1995). *Vital Dust*. BasicBooks, New York, 362pp.

Stephen J. Gould (1996). *Life's Grandeur*. Jonathan Cape, London, 244pp.

Bruce Jakosky (1998). *The Search for Life on Other Planets*. Cambridge University Press, Cambridge, UK; 326pp.

Percival Lowell (1906). *Mars and its Canals*. The Macmillan Company, New York, 393pp.

William Sheehan (1996). *The Planet Mars. A History of Observation and Discovery*. University of Arizona Press, Tucson, 270pp.

Ben Zuckerman & Michael H. Hart (editors) (1995). *Extraterrestrials—Where are They?* Cambridge University Press, Cambridge, 239pp.

The Planetary Report, Mars Underground News, and *Bioastronomy News*. These are regular publications of the Planetary Society, 65 North Catalina Avenue, Pasadena, California 91106–2301, USA: http://planetary.org

Marsbugs. This is a regular electronic newsletter. marsbug@msn.com

And to keep right up-to-date . . .

http://www.nasa.gov
http://cmex-www.arc.nasa.gov/
http://mars.jpl.nasa.gov/
http://www.reston.com/astro/mars/
http://photojournal.jpl.nasa.gov/   A complete set of NASA's images
http://www.science.org.au/nova/sponsors.htm

THE FUTURE

. . . as seen by a former astronaut . . .

Michael Collins (1990). *Mission to Mars.* Grove Weidenfeld, New York, 307pp.

. . . as seen by an engineer and insider in the space program . . .

Robert Zubrin (1996). *The Case for Mars.* The Free Press, New York, 328pp.

. . . and as seen by science fiction writers . . .

Ben Bova (1993). *Mars.* Spectra
Kim Stanley Robinson (1992, 1993, 1996). *Red Mars. Green Mars. Blue Mars.* HarperCollins Publishers, London.

A TASTE OF THE PRIMARY SCIENCE

Stefan Bengtson (editor) (1994). *Early Life on Earth.* Nobel Symposium 84, Columbia University Press, 630pp.
Gregory Bock and Jamie Goode (editors) (1996). *Evolution of Hydrothermal Ecosystems on Earth (and Mars?).* Wiley, Chichester (Ciba Foundation Symposium 202), 334pp.

Michael Carr (1996). *Water on Mars*. Oxford University Press, Oxford, 229pp.

Paul J. Thomas, Christopher F. Chyba & Christopher P. McKay (editors) (1997). *Comets and the Origin and Evolution of Life*. Springer-Verlag, New York, 296pp.

ANTARCTICA AS AN ANALOGUE

Tim Bowden (1997). *The Silence Calling*. Allen & Unwin, Sydney, 593pp.

LESSONS OF THE PAST

Peter Whitfield (1998). *New Found Lands—Maps in the History of Exploration*. Routledge, New York, 200pp.

WHY BOTHER?

Preston Cloud (1978). *Cosmos, Earth and Man, a Short History of the Universe*. Yale University Press, New Haven, 372pp.

Pontifical Academy of Science (1997). *Plenary Session on the Origin and Early Evolution of Life (Part I); Reflection on Science at the Dawn of the Third Millennium (Part II); Round Table on the Problems of the Origin of Life (Round Table)*. October 1996. Commentarii IV (3), Vatican City.

Carl Sagan (1994). *Pale Blue Dot. A Vision of the Human Future in Space*. Random House, New York, 427pp.

Stuart Ross Taylor (1999). *Destiny or Chance: Our Solar System and Its Place in the Cosmos*. Cambridge University Press, Cambridge, UK, 256pp.

INDEX

Index compiled by Russell Brooks

165

167